MISSION: MARS

MISSION: MARS

THE THRILLING HISTORY OF THE RED PLANET

JOHN C. HAMILTON

RavenFire
Media, Inc.

RavenFire Media, Inc.

Last updated November 2022.

ISBN: 978-0-9828459-7-4

www.johnchamilton.com

CONTENTS

MARS MYTHS AND LEGENDS

1. Mysterious Mars 3
2. Tracing the Stars and Planets 5
3. Mars in Greek and Roman Mythology 9
4. Schiaparelli and the Martian Canals 11
5. The War of the Worlds 15
6. The Panic Broadcast 19
7. The Face of Cydonia 21
8. Mars in Popular Culture 24

MARS: THE RED PLANET

9. The Red Planet 29
10. Maps of Mars 30
11. Ancient History 33
12. Orbit 34
13. Size and Gravity 36
14. Why Is Mars Red? 38
15. Atmosphere 39
16. Climate and Weather 41
17. Geology 44
18. The Grand Canyon of Mars 46
19. Olympus Mons 48
20. Poles 50
21. Craters 52
22. Water On Mars 54
23. The Moons of Mars 57
24. Life On Mars? 59
25. Exploring Mars 60

MARS ORBITERS

26. Journey to the Red Planet 65
27. Mariner 4 67

28. Mariners 6 and 7 69
29. Mariner 9 71
30. Mars 2 and 3 74
31. Viking 1 and 2 76
32. Mars Global Surveyor 80
33. 2001 Mars Odyssey 82
34. Mars Express 85
35. Mars Reconnaissance Orbiter 88
36. Mangalyaan 91
37. Maven 93
38. CubeSats 95
39. ExoMars 97
40. July 2020 Launch Window and the Future 99

MARS LANDERS

41. Landing On Mars 103
42. Why Go To Mars? 104
43. Mars 2 and 3 106
44. The Viking Landers 108
45. Touchdown! 112
46. Mars Pathfinder 116
47. Mars Polar Lander 119
48. Beagle 2 121
49. Phoenix Mars Lander 123
50. Schiaparelli EDM Lander 126
51. InSight 128

MARS ROVERS

52. Journey Across An Alien World 135
53. Sojourner 136
54. Pathfinder and Sojourner Reach Mars 139
55. The First Rover To Explore Mars 142
56. Spirit and Opportunity 145
57. Craters and Dust Devils 148
58. The Long Journey 151
59. Curiosity 155
60. Seven Minutes of Terror 158
61. Exploring Gale Crater 161
62. Perseverance 165

63. Zhurong 170
64. Future Rovers 172

HUMANS TO MARS

65. The Lure of the Red Planet 175
66. Can Humans Live On Mars? 177
67. Why Should We Go? 178
68. Training For Mars 180
69. Getting To Mars 182
70. The Orion Spacecraft 184
71. Space Launch System 186
72. The Long Voyage to Mars 189
73. Landing 191
74. Living on Mars 193
75. A Base On Mars 195
76. How Space Affects the Body 198
77. Spacesuits 201
78. Finding Water 203
79. Growing Food 204
80. Making Fuel 206
81. Terraforming Mars 208

Also in the Destination Outer Space Series 211
Get a Free Book 213
Please Leave a Review 215
Timeline 217
Glossary 223
Selected Bibliography 229
About the Author 235
Also by John C. Hamilton 237

MARS MYTHS AND LEGENDS

CHAPTER 1
MYSTERIOUS MARS

For much of our history, the planet Mars has represented death and chaos. To the ancient Greeks and Romans, Mars was the god of war, the destroyer. People in early civilizations all over the world looked skyward to that rust-colored point in the sky and trembled. What is it about Mars that frightened them? Even today, when "space invaders" are mentioned, we think of Martians. Invaders from Pluto or Venus just don't seem to terrify us as much. But mention Martians, and you stop a moment, eyes widening maybe just a bit, before you chuckle nervously and move on.

Maybe it's because Mars appears so similar to our own Earth. Both planets have polar ice caps, seasons, and a 24-hour day. It's natural to think there might be some sort of life on Mars. Yet, ancient civilizations knew none of this—to them, Mars was a red-hued light in the sky. Why were they so scared?

Mars, the God of War.

Mars is part of our culture, our mythology. It has been for thousands of years. Today, in our scientific world, we believe everything can be explained. If there's a mystery, we experiment. We analyze the contents of a test tube or peer through a powerful telescope until the problem is solved. But it wasn't always so.

Even before the age of science, people searched for answers to the riddles of nature and human behavior. They turned to myths. The Latin origin of the word "myth" is *mythos*, a story. Myths give life to society's hopes, dreams, and fears.

Myths persist. They're often hidden, but they seep into our unconscious minds. In *The Power of Myth*, author Joseph Campbell said mythology is "the song of the universe." Myths help us understand characters—called archetypes—such as heroes, villains, shapeshifters, and tricksters.

But what about Mars? How did the Mars myth come to symbolize death and destruction? Why is Mars so frightening? The explanation is there for everyone to see. Venture outside on a clear night and gaze upward. You can find the answer in the stars.

CHAPTER 2
TRACING THE STARS AND PLANETS

THOUSANDS OF YEARS AGO, ASTRONOMERS OBSERVED THE STARS IN order to understand our place in the universe. The movement of the heavens was reassuringly predictable. Stars invariably rose in the east and set in the west, just like the Sun and Moon. They were arranged in patterns that resembled creatures and objects. We call them constellations today.

The positions of the stars shifted as the seasons changed, but relative to each other, the patterns stayed the same. Sometimes strange lights would flash across the horizon (meteorites), or bright balls with eerie glowing tails hung in the sky for many nights (comets). But these events came and went. The stars always remained.

The Greek words meaning "law of the stars" gave us the word astronomy. With the right record keeping, the sky can be used like a calendar. By tracking the predictable star patterns, early astronomers helped people choose the best time to plant and harvest crops, or decide when seasonal herds of animals would pass through their lands. In fact, every advanced civilization of the past, from China to Central America, developed some sort of astronomical knowledge. Most also had religious beliefs connected to their understanding of the heavens.

Many civilizations invented devices, or built structures, to help their astronomers track the stars. In Egypt, the Great Pyramid of Giza is aligned with the polestar (Polaris, the North Star). Seasons are tracked by the position of the pyramid's shadow. A temple constructed by the ancient Anasazi people in Chaco Canyon, New Mexico, has a window through which a shaft of sunlight enters and covers a special spot on the opposite wall only on June 21, the summer solstice (the longest day of the year). Stonehenge, a circular formation of enormous stone slabs in England, was built about 4,500 years ago. It accurately predicted the positions of the Sun and Moon, and the changing seasons.

The Sumerians, the Babylonians, Egyptians, as well as the Mayans and Aztecs of Central America, all developed accurate calendars and almanacs based on the regular movement of the stars. Navigation aids were also invented that relied on predictable star patterns.

However, five lights in the night sky acted strangely. Today, we know that these are the five planets that can be seen without a telescope: Mercury, Venus, Mars, Jupiter, and Saturn. To ancient people, these "stars" wandered through the sky. One night a star might be near the constellation of Scorpius, but a few weeks later shift positions to Sagittarius or Capricornus. (In fact, the word "planet" comes from the Greek planetes, which means "wanderers.") Sometimes, these strange lights even moved backward for a short time, or looped, before continuing their eastward movement through the string of constellations (the zodiac). Astronomers call this backward movement retrograde motion.

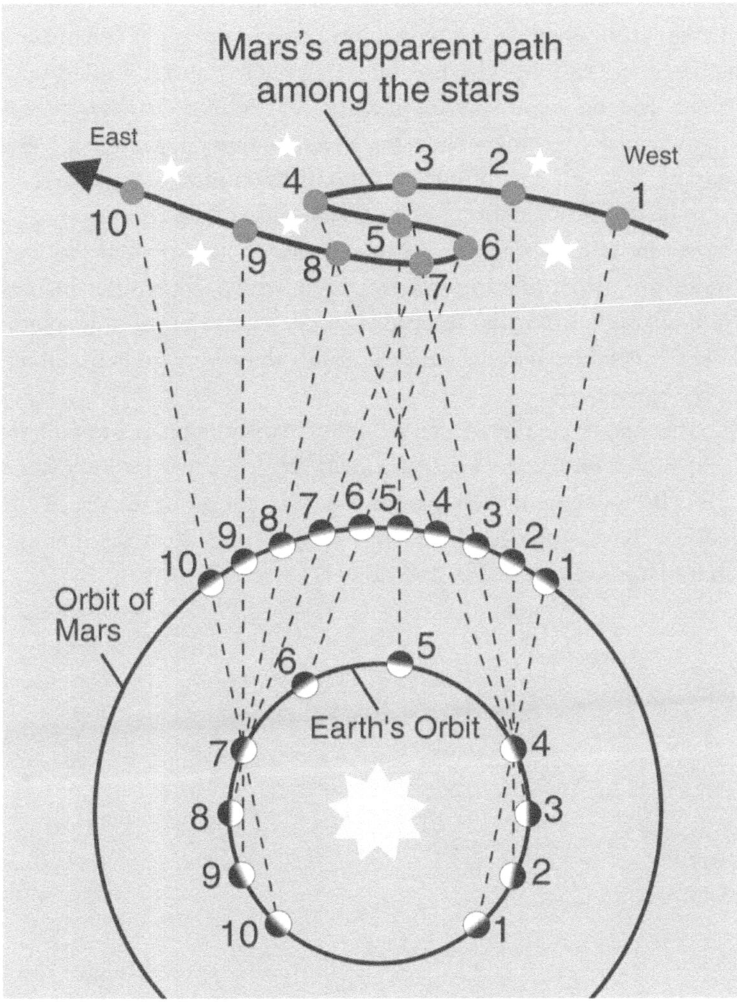

This chart shows the retrograde motion of Mars. Because Earth moves faster around the Sun, Mars appears to backtrack (points 4, 5, and 6) before resuming its west-to-east movement through the zodiac.

Predictable things make us feel safe, like we're in control. We grow fearful when an unpredictable event happens. When ancient astronomers saw wandering stars, they were naturally filled with dread. Maybe some sort of disaster was looming. (*Disaster* is a Greek word meaning "bad star.") Astronomers could track the planets through the night sky, but they couldn't explain their complicated movements.

When there is a gap between our understanding and the way things are, myth fills the space. That is why we gave the planets human traits. It was an attempt to make sense of the unexplainable. The planets became supernatural—they were gods that lived in the heavens.

There was one planet that troubled astronomers most of all. Its light never flickered like a star, and it glowed an ominous red color. Its movement through the sky was the most erratic of the planets. It was the least predictable, and the most frightening. That planet was Mars, the god of war.

CHAPTER 3
MARS IN GREEK AND ROMAN MYTHOLOGY

About 3,000 years ago, the ancient Greeks believed in many gods. The 12 Olympians were the most powerful. They were the controllers of natural forces such as the wind, the oceans, and earthquakes. The five known planets were each named after different gods, depending on how they moved in the heavens. For example, Mercury, having the swiftest path through the zodiac, was named after Hermes, the quick messenger of the gods. Mars was named Ares, the god of war, because of its unpredictable, unusual motion. Later, during the ancient Roman period, the planet was called as we know it today—Mars.

Much of ancient Greek culture was passed down to modern Western civilization. Like the ancient Greeks, today we value reason and logic, fairness, and a curiosity of the natural world. The opposite of the Greek ideal is war—fighting, murder, and disorder. Mars, the planet, represented all these terrible traits as Ares, the god of war.

Greek cults sprang up in which Ares was worshipped, especially in the city-state of Sparta. Later, in ancient Rome, Ares became known as Mars. The warrior god's cult became even stronger, with Roman citizens believing that Mars was their protector. Ancient Rome was a society built on conquest, where war and conflict were common. And as the Romans invaded other lands, so spread the cult of Mars.

Myths are powerful forces in civilizations. They teach us how to behave in society, or how not to behave. Our myths are passed down from generation to generation until their messages are assumed to be "the truth." We believe them without question. That's why, even to this day, when we think of Mars, we imagine negative things that our society fears and hates. Because of our Greek heritage, Mars will always be looked on with suspicion and fear.

CHAPTER 4
SCHIAPARELLI AND THE MARTIAN CANALS

During the late 1800s, better telescopes opened the skies to fresh observations. It was during this time that astronomers discovered that Earth's Moon was a dead, crater-filled place. There was no life to be found on our nearest neighbor.

If the Moon was barren, then the next best place to look for extraterrestrial life might be the planet Mars. But even with the largest, most advanced telescopes available, the Red Planet was difficult to observe. Faint polar caps of ice could be seen. Large splotches of blue-green color were mixed with the reddish Martian soil. Astronomers wondered if these areas could represent oceans, or maybe even vegetation, since they seemed to change size with the seasons.

Astronomers kept their instruments trained on Mars, searching for more signs of life. In 1877, the orbits of Mars and Earth brought them closer together than they had been in many years, making for very clear observations. That year, American astronomer Asaph Hall, using the 26-inch (66-cm) telescope at the U.S. Naval Observatory in Washington, DC, discovered the existence of Mars's two moons, Phobos and Deimos.

Astronomer Giovanni Schiaparelli was the director of the

Brera Observatory in Milan, Italy. During his long career, he made many discoveries and observations, especially of comets and meteors. But in 1877, with his 8-inch (20-cm) telescope focused on Mars, Schiaparelli announced a finding that changed the way we think of the Red Planet.

Giovanni Schiaparelli

Schiaparelli had very sharp eyesight. It was a valuable trait for astronomers during this time, before photography was widely used with telescopes. Using his superior vision, Schiaparelli drew maps of Mars that included what he thought were oceans and continents. Then, in 1877, Schiaparelli saw enormous lines crisscrossing the planet.

At first, the astronomer doubted his own discovery. Mars was hard to see, even with the best telescopes. But night after night, Schiaparelli saw these lines, which he called *canali.* It is an Italian word that means "channels" or "grooves."

In the 1960s and 1970s, when NASA sent its Mariner missions to Mars, scientists discovered a 3,000-mile (4,828-km) -long canyon called Valles Marineris. Did Schiaparelli see Valles Marineris all the way back in 1877? We may never know. Some scientists today explain his observations as optical illusions. But whatever he saw, his announcement caused a worldwide sensation.

Many people misunderstood Schiaparelli's use of the Italian term *canali.* They thought the astronomer had observed "canals" on Mars. It was proof, they believed, that there was intelligent life on the Red Planet. Some said that the Martians must have dug the canals to bring melting water from the polar ice caps to irrigate parched regions of the thirsty desert world.

A few years before Schiaparelli's discovery, in 1869, the Suez Canal had been completed in Egypt. It was a technological marvel for its time and only proved to people that these canal-building Martians must have had a very advanced civilization indeed. Schiaparelli was more cautious about his discoveries. He believed there was probably a natural explanation for the lines he was seeing on Mars. He kept searching for more clues until his failing eyesight forced him to stop in 1892.

Percival Lowell

The news of the Martian "canals" caused many people to point their telescopes toward the Red Planet. Percival Lowell, who in 1894 built a large observatory on a mountain near Flagstaff, Arizona, was certain that the canals were real. He spent many years observing Mars and searching for life. He wrote three popular books on the subject. His enthusiasm fueled the public's imagination.

This popular belief in life on Mars led to an explosion of stories about the Red Planet. In these early science-fiction tales, authors tried to guess what kind of superior beings lived on Mars. Most agreed they had to be smarter and wiser than humans.

Then, in 1897, British author H.G. Wells published a novel that forever changed how we think of Martians.

CHAPTER 5
THE WAR OF THE WORLDS

HERBERT GEORGE (H.G.) WELLS WAS AN ENGLISH AUTHOR. HE WAS most famous for his science-fiction novels. He wrote over 80 books, including *The Time Machine* in 1895, and *The Invisible Man* in 1897.

During the end of the 1800s, people were alarmed that Germany was strengthening its army and navy. Many countries feared a German invasion. Wells used this fear, plus the horror of modern weapons and technology, as themes in his new novel, *The War of the Worlds*.

H.G. Wells

The story was first serialized in *Pearson's Magazine* in 1897. It was then published as a novel in 1898. *The War of the Worlds* is a classic alien invasion story. Martians flee their dying world and land on Earth in giant cylinder-shaped rockets. Using enormous "fighting machines" (pods on long metal legs), they begin their invasion by attacking London, England.

Their weapons are "heat rays" and toxic black smoke. The citizens of Earth are helpless against the Martian technology. Then, the aliens drop dead. They had no natural defense against the simplest of Earth creatures: germs.

A Martian fighting machine attacks in an illustrated 1906 edition of H.G. Wells's *The War of the Worlds*.

The War of the Worlds begins ominously. An unknown threat appears to be coming from Mars. Following is a section of the novel's famous opening paragraph:

"No one would have believed in the last years of the nineteenth century that this world was being watched keenly and closely by intelligences greater than man's... that as men busied themselves about their various concerns they were scrutinized and studied, perhaps almost as

narrowly as a man with a microscope might scrutinize creatures that swarm and multiply in a drop of water... No one gave a thought to the older worlds of space as sources of human danger... Yet across the gulf of space, minds that are to our minds as ours are to those of the beasts that perish, intellects vast and cool and unsympathetic regarded this Earth with envious eyes, and slowly and surely drew their plans against us."

When H.G. Wells created *The War of the Worlds,* he did more than write an alien invasion story. He also criticized British society. At the time Wells wrote the novel, Great Britain was busy invading countries all over the world and turning them into colonies. Some people compare the events in *The War of the Worlds* to Europe's invasion of Tasmania. Great Britain began settling the island in the late 1700s. In the years that followed, many native Tasmanians were killed in warfare or died from diseases brought by the Europeans. Tasmania today is a state of Australia.

The War of the Worlds was also about one of Wells's favorite themes—survival of the fittest. A clear example happens at the end of the novel, when the seemingly invincible invaders are killed off by simple Earth bacteria.

The War of the Worlds was a tremendous success for H.G. Wells. It came out during a time of great public interest in Mars. The book had a big impact on our culture; alien invasion stories remain popular to this day.

For 30 years after the publication of *The War of the Worlds,* writers fed the public's imagination by churning out tales of the Red Planet by the hundreds. Edgar Rice Burroughs, author of the Tarzan of the Apes stories, wrote a series of popular novels about Mars (called "Barsoom"), starting in 1917 with *A Princess of Mars.* These books told of the adventures of John Carter, a swashbuckling hero who overcame Martian warriors with the use of sword and magic.

In 1938, there was a fictional story that had an even bigger impact on how we look at Mars, and it didn't come on the printed page. It was told on a relatively new medium, one with an untested power to sway the public imagination: radio.

A Princess of Mars
EDGAR
RICE
BURROUGHS

CHAPTER 6
THE PANIC BROADCAST

On October 30, 1938, Halloween eve, America was treated to a prank. That night, a CBS Radio play based on H.G. Wells's *The War of the Worlds* was broadcast. Written by Howard Koch, it starred 23-year-old director and actor Orson Welles and The Mercury Theatre on the Air.

Orson Welles

The play began with a mock broadcast from a hotel in New York City. It featured a live performance by Ramon Raquello and his orchestra, but the music was soon interrupted by news reports of Martians landing in a small New Jersey town called Grovers Mill. The invaders soon opened fire with their "heat rays." The Martians destroyed everything in their path, including New York City. Some of the fake news reports were modeled after the famous 1937 Hindenburg airship disaster broadcast. Combining realistic sound effects and riveting narration, *The War of the Worlds* radio play thrilled millions.

In fact, for many people, it was *too* thrilling. In 1938, most Americans were hearing daily reports of violence in Latin

America and Asia, and Nazi Germany was threatening much of Europe. The public was already suffering from "war jitters," and it didn't seem so far-fetched that invaders from Mars could land on American soil.

William Dock defends his farm against the Martians near Grovers Mill, New Jersey.

The night *The War of the Worlds* was aired, panic swept the nervous nation. Many people tuned in their radios too late to hear Welles introduce the fictional play. As many as one million people actually believed Earth was being invaded by Martians. Some terrified citizens rushed into the streets, or even fled the cities. Local legend says a shotgun-wielding farmer shot at a water tower in Grovers Mill, mistaking it that dark night for a Martian fighting machine.

The panic broadcast doesn't make much sense to us today. The whole play lasted just 45 minutes, far too short a time to blast off from Mars, travel 35 million miles (56 million km) through space, and then conquer the entire Earth. That so many people were fooled proved how powerful the new technologies of radio and other mass media could be.

Orson Welles was stunned by the public's reaction. He later told reporters he had been unsure whether to perform the play at all. "It was our thought," he said, "that people might be bored or annoyed at hearing a tale so improbable."

At the start of The Mercury Theatre on the Air's next regular broadcast, Welles apologized for the hysteria he had caused. But the play made him famous. He went on to Hollywood to direct film classics such as *Citizen Kane* and *The Magnificent Ambersons.* Still, it was the panic broadcast of 1938 for which many will remember Welles.

CHAPTER 7
THE FACE OF CYDONIA

DURING THE SUMMER OF 1976, THE VIKING 1 ORBITER WAS BUSY taking photographs of the surface of Mars. During one of its orbits, the spacecraft passed over the flat plains of Cydonia, a region in the northern hemisphere of the planet. NASA scientists were surprised when they saw an image beamed to Earth. Staring back at the camera was a huge, humanoid face. It had a nose, eyes, and even a mouth.

The "Face" on Mars is an eroded rock formation. It measures about 1.5 miles (2.4 km) long, 1.25 miles (2.0 km) wide, and .25 miles (.4 km) tall. When it was first discovered, experts dismissed it as a trick of the lighting. Shadows in the rocks give the illusion of facial features.

The Face was not the only unusual feature seen in Cydonia. Several other curious structures were found in the same area. They included what appeared to be pyramids. There was also a formation that looked like the ruins of an ancient city, or fortress.

Scientists were sure these structures, including the Face, were unusually shaped rocks and hills, like mesas in the American Southwest. When they released the photo to the public, they thought it would be an interesting curiosity. Little did they realize what a sensation it would cause.

Starting in 1977, the Face of Cydonia became a hot topic. Many people were certain that it proved the existence of a past civilization on Mars. Several books presented wild theories about its origins. Some believed a long-dead race of Martians left the Face behind. It was a monument to their society, just like the pyramids of the ancient Egyptians. Others were convinced NASA was covering up the truth about the Face.

Scientists said all the conspiracy theories were nonsense. The Face, the pyramids, and the fortress are simple rock formations. The Face looks like a human because of an optical illusion, a coincidence in lighting and natural erosion. We know that on Mars there is wind and seismic activity (marsquakes). We also know that in the planet's ancient past, running water eroded the landscape.

NASA's Mars Global Surveyor spacecraft arrived in Mars orbit in 1997. The probe took photos of the Martian surface that were 10 times sharper than the Viking 1 orbiter of 1976. When scientists received the new images, they saw… rock formations, just as they had predicted. There was not a single alien in sight.

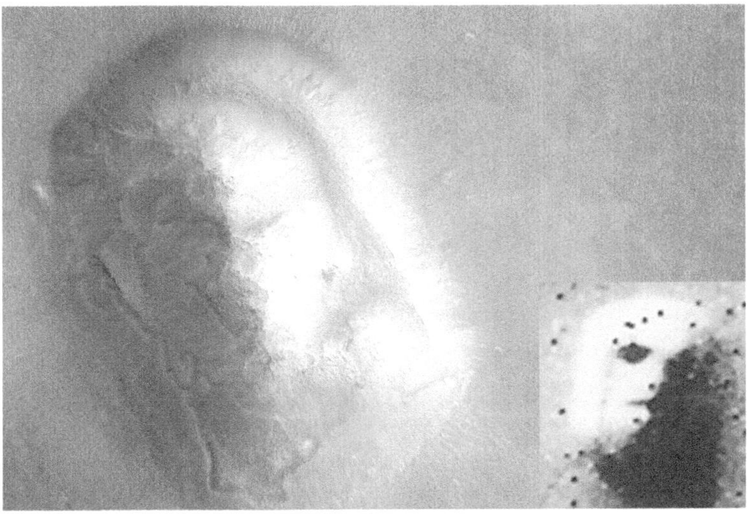

The infamous Face of Cydonia looks very different in this photo taken by the Mars Reconnaissance Orbiter in 2007. The better-quality photo shows the eroded Mars mesa in more detail.

CHAPTER 8
MARS IN POPULAR CULTURE

IN THE MID-1900S, AUTHORS CONTINUED TO WRITE STORIES ABOUT Mars. Ray Bradbury gave the subject a new spin with his masterpiece, *The Martian Chronicles.* First published in 1951, it is a collection of short stories that record the human colonization of Mars. Instead of Martians conquering Earth, it's humans who invade the Red Planet.

In the book, Martians have died out, but somehow continue to haunt the human visitors. The aliens in *The Martian Chronicles* are wise and peaceful, unlike many of the humans who settle on the planet. In the end, it's the Earthlings who become the new Martians.

In *The Sands of Mars* (1951), by Arthur C. Clarke, a writer visits a colony on the Red Planet and discovers native life forms. Another science-fiction classic about Mars is Robert Heinlein's *Stranger in a Strange Land.* It won the Hugo Award in 1962.

Starting in the 1960s, a series of NASA Mariner spacecraft proved Mars was a cratered, desert planet, hostile to life. Writers began crafting stories that showed this new reality. Many tales were about Earthlings colonizing and terraforming Mars. Some of the most popular modern books include *The Martian Race* (1999) by Gregory Benford; *Mars* (1992) by Ben Bova; and *Red Mars* (1993) by Kim Stanley Robinson (plus its two sequels, *Green Mars* (1993) and *Blue Mars* (1996)). In 2011, author Andy Weir published *The Martian.* It has sold over five million copies, making it one of the most successful books about Mars in recent times.

People have always had an enormous appetite for movies and television shows about the Red Planet. Dozens of films have been made about Mars or Martians since the 1950s. Notable examples include *Invaders From Mars* (1953), *War of the Worlds* (both the 1953 version and Steven Spielberg's 2005 film), *Total Recall* (1990) starring Arnold Schwarzenegger, and *Mars Attacks!* (1996) by director Tim Burton. Some movies were made strictly for laughs, including Disney's *RocketMan* (1997). Even Bugs Bunny got in on the act. In 1948, Looney Tunes introduced Marvin the Martian. With his Roman centurion's helmet, deadly ray gun, and plans to blow up Earth, Marvin was a worthy opponent for that "wascally wabbit" Bugs.

The Expanse is a hit television series based on the Hugo Award-nominated books by James S. A. Corey. It takes place 200 years in the future, when people from Earth have colonized Mars and the nearby asteroid belt. A delicate peace is shattered as each side tries to dominate the solar system and the galaxy beyond.

In 2016, the National Geographic Channel began a docudrama called *Mars*. The miniseries combines interviews with real-life experts with a fictional story that takes place in the year 2033. In season one, the first astronauts to set foot on Mars accidentally land far from their base. They must grapple with the harsh Martian landscape as they journey to safety.

In 2015, director Ridley Scott brought Andy Weir's 2011 novel *The Martian* to life on the big screen. Like the book, the movie followed astronaut Mark Watney's struggle to survive being stranded on Mars. Both the book and film are hailed for their devotion to realism. Despite his grim situation, Watney retains his humor, and uses his knowledge of science to make his survival and rescue possible. The outdoor scenes in *The Martian* were filmed at Wadi Rum, in southern Jordan, because the red sand and stark rock outcroppings closely resemble the landscapes found on Mars.

Andy Weir's *The Martian* has sold more than five million copies. It was made into a film in 2015 starring Matt Damon.

The power of myth to influence our feelings about Mars has been affecting our culture for thousands of years. Perhaps someday, when humans are firmly established on Mars, we'll lose our fear and dread of the Red Planet. But in the meantime, the myth lives on.

As Orson Welles said at the end of *The War of the Worlds* panic broadcast in 1938, "So good-bye everybody, and remember, please,... the terrible lesson you learned tonight. That grinning, glowing, globular invader of your living room is an inhabitant of the pumpkin patch, and if your doorbell rings and nobody's there, that was no Martian... it's Hallowe'en."

MARS: THE RED PLANET

CHAPTER 9
THE RED PLANET

FOR THOUSANDS OF YEARS, PEOPLE HAVE STARED UP AT THE PLANET Mars and marveled at its mysteries. The blood-red disk in the night sky captured the imagination of writers, poets, and philosophers. The ancient Greeks and Romans believed Mars was the god of war. Its fierce reputation continues to this day. The Red Planet has been the subject of countless science fiction books, stories, and movies. They include *War of the Worlds*, *The Expanse*, and *The Martian*, to name just a few.

Today, Mars is much more than a frightening object of imagination. It is an actual place that we have explored not only with powerful telescopes, but also a fleet of orbiters, landers, and rovers. In just the past two decades, scientists have learned more about Mars than in all of human history combined.

Mars continues to surprise and inspire. With manned missions planned for the not-too-distant future, we will certainly discover more about the mysterious Red Planet. Perhaps people will eventually live there in permanent settlements. The science we learn on faraway Mars may even someday help us repair and sustain our own planet Earth. The challenges are many, but the possibilities are endless.

CHAPTER 10
MAPS OF MARS

THE FOLLOWING MAPS SHOW THE WESTERN AND EASTERN HEMISPHERES of Mars. They were created by combining thousands of photos taken by NASA's Mars Global Surveyor, an orbiting satellite that mapped Mars from 1999 to 2006. The maps show the true colors that astronauts would see when traveling to Mars.

- Mons, montes: mountain
- Planitia, planitiae: low plain
- Planum, plana: plateau or high plains
- Terra, terrae: extensive area
- Vallis, valles: valley
- Vastitas, vastitates: extensive plain

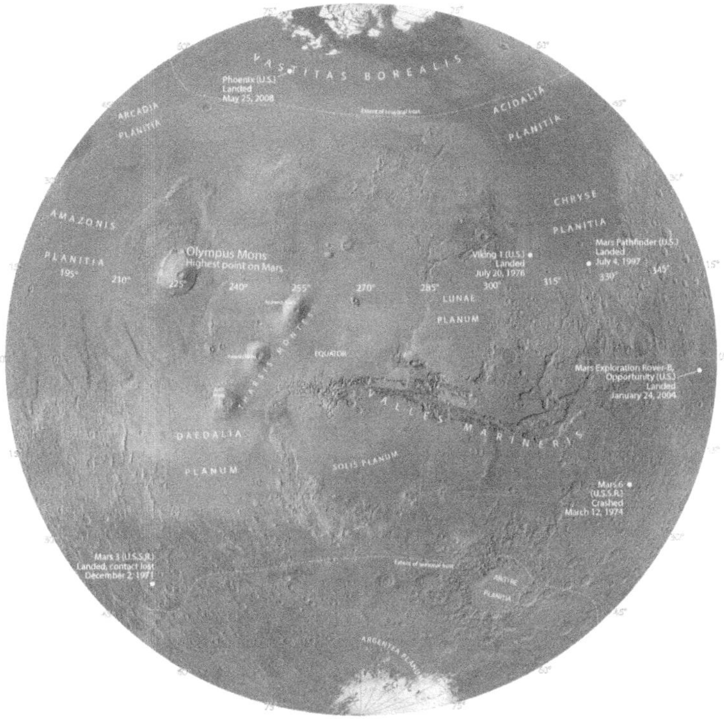

Mars Western Hemisphere

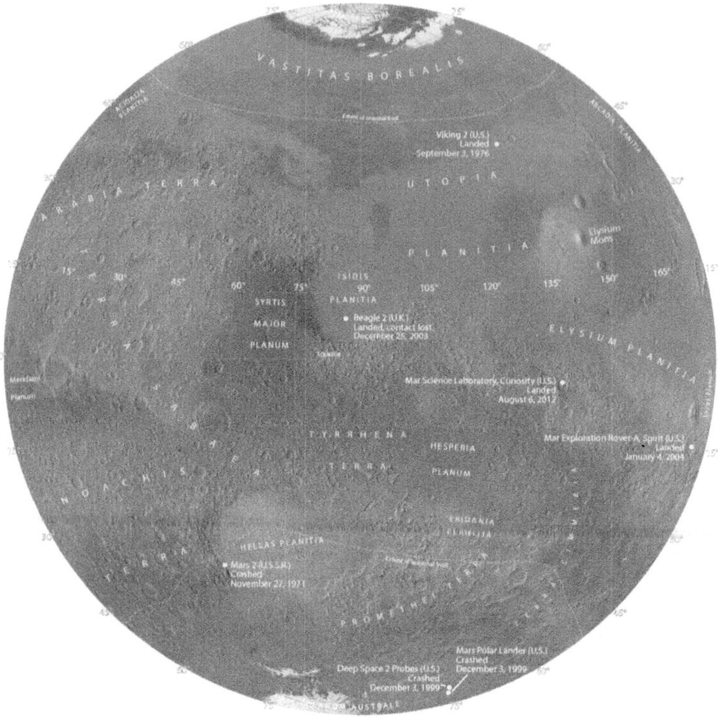

Mars Eastern Hemisphere

CHAPTER 11
ANCIENT HISTORY

OUR SOLAR SYSTEM FORMED FROM A COLLAPSING CLOUD OF interstellar dust and gasses 4.6 billion years ago. About 3.5 billion years ago, Mars and Earth were much more similar than they are today. Both had mild temperatures and oceans of water. We know this because there are geological features on Mars that could only have been formed by water. They include dry riverbeds, deltas, lakes, and ocean basins. There are also layers of sedimentary rocks that need water to form.

Water is a central ingredient for life to arise. It supports the complex chemical reactions that living things depend on. By the time life began on Earth, it is possible that some sort of living organisms may have already existed on Mars.

Over billions of years, the atmosphere on Mars became thinner and thinner. Eventually, all the surface water evaporated or froze. There was too little atmospheric pressure for liquid water to exist. Today, Mars is a cold, dry place, too harsh for humans to live there unprotected.

CHAPTER 12
ORBIT

MARS IS THE FOURTH PLANET FROM THE SUN, THE OUTERMOST OF THE four terrestrial planets. The terrestrial planets are rocky and similar in size to Earth. They are also closest to the Sun. The four outer planets of the solar system are gas and ice giants that are much larger than Earth.

A day on Mars (a sol) lasts about 24 hours and 39 minutes. That is not much longer than a day on Earth. Mars orbits the Sun every 687 Earth days. A 15-year-old on Earth would have had only 8 birthdays on Mars.

Mars is about 142 million miles (229 million km) from the Sun. By comparison, Earth is 93 million miles (150 million km) from the Sun. The speed of light is 186,000 miles per second (299,338 km/sec). It takes light from the Sun almost 13 minutes to reach Mars. Its greater distance from the Sun is one reason Mars is so much colder than Earth.

The distance between Mars and Earth is always changing as the planets orbit the Sun. At their closest (perihelion), they are about 35 million miles (56 million km) apart. That is near enough for Mars to appear as a bright red object in the night sky. When Earth and Mars are on opposite sides of the Sun (aphelion), they are about 250 million miles (402 million km) distant from each other. About every 26 months, the planets are lined up properly for spacecraft to travel to Mars. A journey to the Red Planet takes about six to eight months, using current rocket technology.

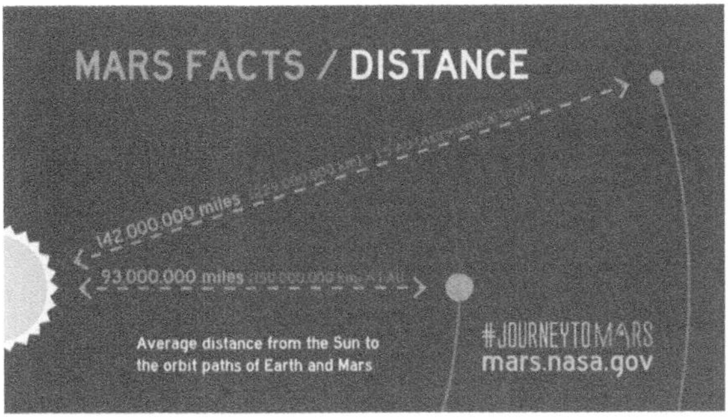

CHAPTER 13
SIZE AND GRAVITY

MARS IS ABOUT HALF THE SIZE OF EARTH. ITS DIAMETER (THE distance through the planet's center from one side to another) is about 4,220 miles (6,791 km) at its equator. In comparison, Earth's diameter is 7,926 miles (12,756 km).

Mars is not a perfect sphere. Just like Earth, it has a bulge around its equator because it spins, or rotates, around its axis. If measured through the poles, Mars's diameter is slightly shorter, measuring 4,196 miles (6,753 km).

The circumference of Mars (the distance around the equator) is 13,233 miles (21,296 km). The circumference of Earth is 24,874 miles (40,031 km).

Mercury is the only planet in our solar system smaller than Mars. (Pluto is a dwarf planet that is smaller than Earth's Moon.) Mars is smaller than Earth, but it has no liquid oceans. Because of this, its land area is about the same as all the dry land on Earth.

The gravity on Mars is about 38 percent less than on Earth, thanks to the Red Planet's smaller size and density, or mass. A 200-pound (91-kg) astronaut would weigh just 76 pounds (34 kg). People walking on Mars would feel lighter on their feet, and nearly anybody could dunk a basketball!

The terrestrial planets of the solar system, illustrating their comparative sizes. From left to right: Mercury, Venus, Earth, and Mars.

CHAPTER 14
WHY IS MARS RED?

FROM THE INKY BLACK OF THE NIGHT SKY, MARS APPEARS LIKE A glowering red eye staring back at Earth. That is one reason the planet was so feared by early civilizations, and why Mars became known as the god of war by the ancient Greeks and Romans. The truth behind its blood-red hue, however, is far less sinister.

The red color we see from Earth is caused by fine-grained dust. It covers much of the surface of Mars, and in some places is several feet thick. The dust is rich in iron. When exposed to oxygen in the air, the iron rusts (oxidizes) and turns a dull, butterscotch-like color. The oxidized dust is so light that it is easily blown high into the Martian atmosphere. This dust, together with other common rocks on Mars, including grayish volcanic basalt, causes Mars to appear the red color observed from Earth.

CHAPTER 15
ATMOSPHERE

MARS HAS AN ATMOSPHERE, BUT IT IS VERY THIN. ITS AIR IS LESS THAN one percent as dense as the atmosphere on Earth. That is about as rarefied as the air at 30,000 feet (9,144 m) above Earth, which is almost as high as passenger jets fly. With air so thin, hurricane-strength winds on Mars would feel like a spring breeze to an astronaut.

In its younger days, Mars had thick air, much like Earth does today, which allowed liquid water to collect and pool in oceans, rivers, and lakes. Then, over billions of years, the atmosphere began disappearing. The water evaporated, seeped underground, or froze and collected as polar ice.

Most of Mars's remaining atmosphere is carbon dioxide. There are also small amounts of argon, nitrogen, oxygen, carbon monoxide, and water vapor. Traces of methane and ammonia have also been detected.

Because Mars doesn't have a dense atmosphere like Earth, its surface is often struck by asteroids and meteorites. Most space rocks falling to Earth turn into fireballs, or "shooting stars." Friction causes them to burn up as they pass through the thick atmosphere at high speed. On Mars, however, the atmospheric friction is much less, allowing more objects to reach the surface. In addition, Mars is close to the asteroid belt, which means it is a frequent target for space debris captured by the planet's gravity.

Astronomers estimates there are about one million relatively large impact craters on Mars. About 43,000 craters are 5 km (3.1 miles) or more in diameter. About 385,000 are 1 km (.6 mile) or more wide. (On Earth, we are aware of less than 200 impact craters.) Mars's surface is clobbered by space debris so often, in fact, that orbiting satellites with high-resolution cameras, such as NASA's Mars Reconnaissance Orbiter, have even observed new craters appearing on the Martian landscape.

CHAPTER 16
CLIMATE AND WEATHER

MARS IS A BONE-CHILLING PLANET. TO SAY IT IS COLD IS LIKE SAYING a Siberian blizzard is a bit brisk. Typical "warm" Martian days mimic temperatures endured during winters in Minnesota or North Dakota. On average, Martian temperatures are like what you might find in Antarctica. The planet's average surface temperature is a frigid -81 degrees Fahrenheit (-63 degrees C). However, it can sometimes get down to -225 degrees F (-143 degrees C), or briefly climb as high as 70 degrees F (21 degrees C) on sunny summer days near the equator.

Mars is a cold planet for several reasons. It is about one and a half times farther from the Sun than Earth. Also, heat escapes much faster in Mars's thin atmosphere.

Mars is tilted slightly on its axis. As it orbits the Sun, the tilt causes seasons, just like on Earth. The seasons last longer because Mars takes longer to orbit the Sun. During winter, carbon dioxide in the air freezes and collects at the poles. In summer, it evaporates back into the atmosphere. Sometimes frozen carbon dioxide particles form snow, or fog.

Martian winds can be strong enough to produce massive dust storms that cover huge parts of the planet's surface. Dust devils also form, which have been observed by orbiting NASA satellites.

These whirlwinds often create beautiful, swirling designs in the Martian dust.

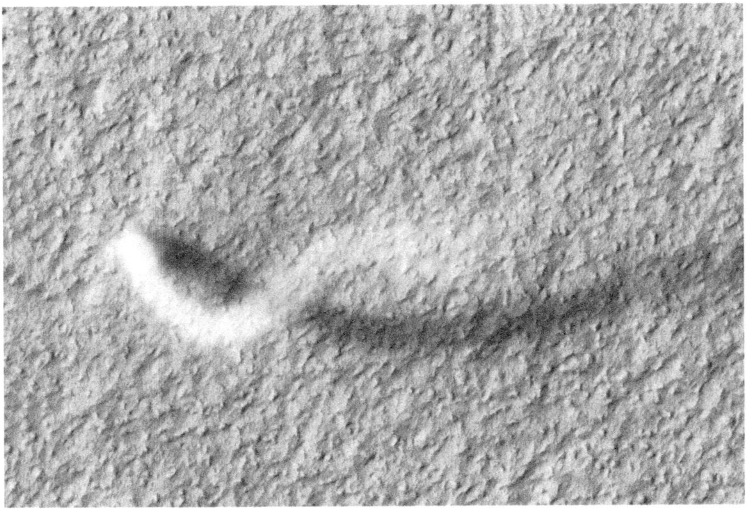

A dust devil swirls across the Amazonis Planitia region of northern Mars. The serpentine winds were photographed from space by NASA's Mars Reconnaissance Orbiter.

CHAPTER 17
GEOLOGY

Mars has a metallic core, much like Earth. Mars's core is made mostly of iron, nickel, and sulfur. The outer part is liquid, composed of molten rock. We know this by studying gravity maps. Like doctors peering at an x-ray, they help scientists "see" inside Mars. The maps were made over a period of 16 years by a trio of NASA satellites orbiting the planet.

Just like Earth, Mars has a soft, paste-like mantle surrounding its core. It also has an outer crust, which averages about 31 miles (50 km) thick. By comparison, the average depth of Earth's crust is 25 miles (40 km).

Earth's crust is broken into gigantic continental plates that rub against each other as they float over the mantle. The areas between the plates experience earthquakes and volcanoes. The crust of Mars isn't broken into plates, although it may have been millions of years ago. Extensive areas of the surface are covered with hardened lava plains. Mars is also home to some of the largest volcanoes in the solar system.

The Martian surface is covered by a layer of talcum powder-like dust containing oxidized (rusted) iron. It is several feet thick in some places. Underneath the dust, Mars's crust consists mostly of volcanic basalt rock. The Martian soil contains nutrients such as sodium, potassium, magnesium, and chloride, which are needed to cultivate plants. Perhaps someday future astronauts will grow crops on Mars.

CHAPTER 18
THE GRAND CANYON OF MARS

MARS IS HOME TO SPECTACULAR SCENIC LANDSCAPES. THAT INCLUDES Valles Marineris, the biggest canyon system on Mars. Dwarfing Earth's Grand Canyon, Valles Marineris is one of the largest canyons in the entire solar system. It is in the southern hemisphere, lying just south of the equator.

Valles Marineris covers almost one-fifth of Mars's circumference. If it were located on Earth, it would stretch across the United States from east to west.

Valles Marineris is about 2,485 miles (4,000 km) long and 124 miles (200 km) wide. In some places, the canyon sinks six miles (10 km) deep. By comparison, the Grand Canyon is 277 miles (446 km) long, 18 miles (29 km) at its widest, and in places slightly more than one mile (1.6 km) deep.

Valles Marineris was discovered in the 1970s when NASA's Mariner 9 probe began orbiting Mars. Scientists have been searching for clues about how the canyon formed. Unlike the Grand Canyon, most of Valles Marineris wasn't created by water erosion. Instead, experts think it was a crack that opened up billions of years ago as large volcanoes caused the nearby Tharsis region to bulge upward. Millions of years later, landslides and rushing floodwaters on Mars may have further widened the giant canyon.

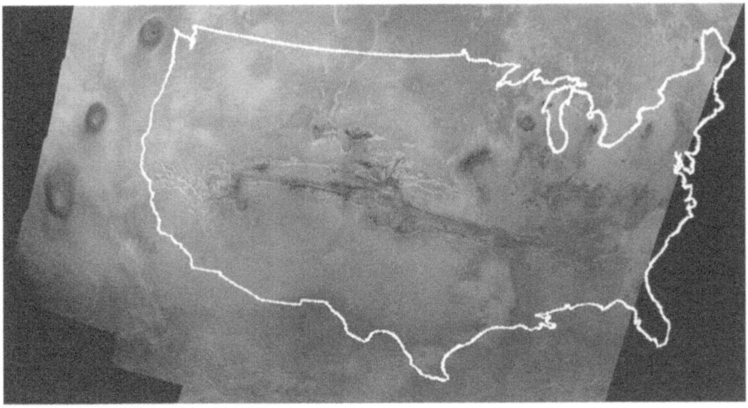

Valles Marineris is big enough to stretch across the continental United States.

CHAPTER 19
OLYMPUS MONS

Olympus Mons is the largest volcano in the solar system. It is an extinct shield volcano, formed almost completely from lava flows. Viewed from the side, shield volcanoes have a low profile, resembling a warrior's protective cover. Hawaii's Mauna Kea is a famous shield volcano. Olympus Mons is about 100 times larger, towering 16 miles (26 km) above the surrounding Martian plains. The massive volcano has a diameter of 374 miles (602 km). That is about the same area as the state of Arizona.

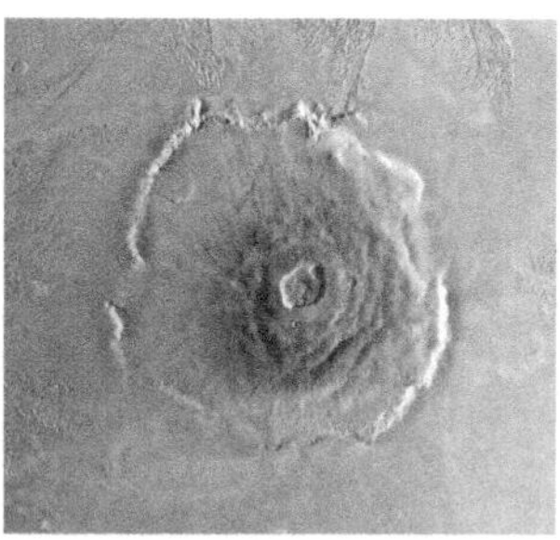

Olympus Mons

Olympus Mons is in the volcanic Tharsis region. Scientists believe magma from the planet's interior pushed Mars's crust upward into a vast plateau. In some spots, magma burst to the surface. Besides Olympus Mons, there are several other colossal volcanoes in this region. The three largest are southeast of Olympus Mons. They include Ascraeus Mons, Pavonis Mons, and Arsia Mons. (*Mons* is a Latin word that means mountain.) Together, they are called the Tharsis Montes. (*Montes* is the Latin plural of mons.)

Why are the volcanoes on Mars so much bigger than those on Earth? There are a few plausible explanations. The eruptions may have been more frequent. The lower gravity on Mars allowed the lava to spread out farther. Also, the Martian planetary crust isn't divided into enormous plates that slowly move, like on Earth. On Mars, once magma surged to the surface, lava kept building up for millions of years in the same spot.

CHAPTER 20
POLES

LIKE EARTH, MARS HAS A NORTH POLE AND A SOUTH POLE. THEY ARE large enough that they can be seen from Earth with a telescope. Each pole has a permanent cap of water ice that is several miles thick. If all the ice were to suddenly melt, it would flood the entire planet with water 33 feet (10 m) deep.

During each pole's winter, constant darkness descends, and the landscape is plunged into severe cold. Carbon dioxide (CO2) in the air freezes and falls to the ground as frost, snow, or dense fog. This "dry ice" is many yards thick. It makes up about 30 percent of the polar ice cap in winter. (The frozen CO2 level at the south pole is much deeper in places.)

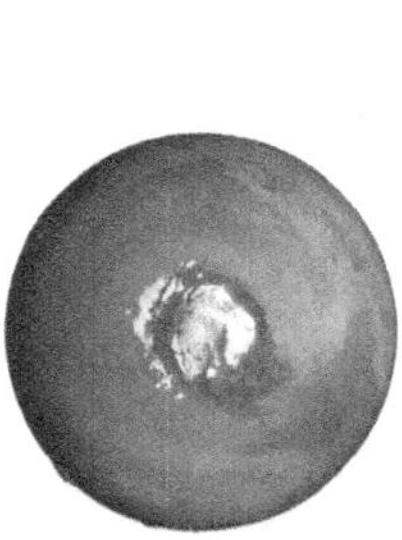

Mars's North Pole

During the warmer, sunnier summer months, the frozen carbon dioxide thaws. It turns back into a gas and returns to the atmosphere. This process generates strong winds, creating vast dust storms. Some water vapor also escapes into the air, which forms wispy clouds that have been observed by NASA satellites. These same spacecraft discovered that when either pole experiences winter, nearly four trillion tons (3.6 million metric tons) of carbon dioxide freezes out of the atmosphere.

CHAPTER 21
CRATERS

THERE ARE HUNDREDS OF THOUSANDS OF CRATERS ON MARS. THEY were caused by passing asteroids, meteoroids, or comets captured by Mars's gravity. When these objects slammed into the Martian surface, they created circular impact craters. Some are small, while others measure hundreds of miles across.

Why does Mars have so many more visible craters than Earth? The atmosphere on Mars is much thinner, which means most space objects do not burn up before reaching the surface. Also, any impact craters that formed on Earth millions of years ago have been smoothed away by wind and rain erosion, or hidden by vegetation.

When you study a map of Mars, you may be surprised to see that most of the northern hemisphere is smooth, covered over by ancient lava flows. The south, on the other hand, is pockmarked by thousands of impact craters. How can this be? One theory is that a gigantic asteroid rivaling the size of Earth's Moon struck Mars billions of years in the past. If true, it would make the northern part of Mars the site of the biggest impact crater ever seen.

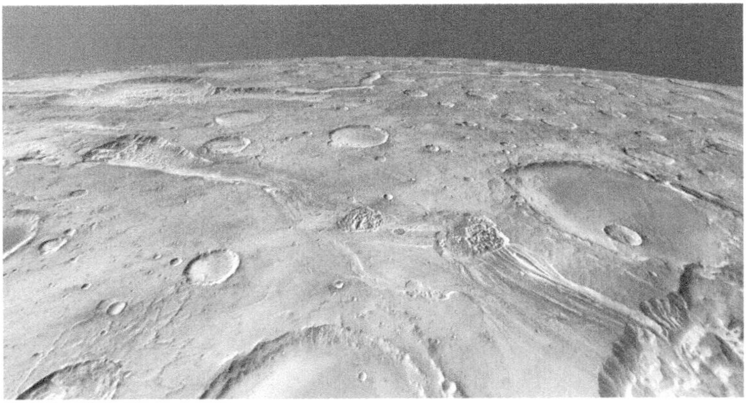

The surface of Mars is heavily cratered.

CHAPTER 22
WATER ON MARS

ABOUT 3.5 BILLION YEARS AGO, OCEANS AND RIVERS OF LIQUID WATER flowed freely on Mars. It was a warmer, wetter world. Evidence includes flow channels carved into the Martian rocks, deltas, lake beds, and river valleys, all of which have been observed by orbiting spacecraft. On the surface, NASA rovers have found rocks and minerals that could only have formed in the presence of water.

In time, something happened to the ancient Martian seas. The atmosphere became too thin, and water couldn't stay liquid any longer. Over billions of years, the water evaporated or froze, until Mars became the dry planet we see today.

Where did the liquid water go? Scientists are still trying to unravel the mystery, but there are tantalizing clues. Billions of years ago, Mars's inner molten core cooled, and the planet lost its protective magnetic shield. For several billion years, energetic particles streaming from the Sun, called the solar wind, eroded Mars's atmosphere. On Earth, our strong magnetic shield protects the atmosphere by deflecting most of the solar wind. On Mars, much of the atmosphere was pushed into outer space. When the atmosphere became too thin, liquid water could no longer exist on the surface. It either froze or evaporated.

Not all the water on Mars was lost to outer space. As the planet's atmosphere slowly thinned over millions of years, much of the water froze and collected at the poles, enough to flood the entire planet many feet deep if it ever melted.

Could some of the water have seeped under the surface? The answer appears to be yes. Using ground-penetrating radar and tests by NASA rovers, many scientists have concluded that a vast amount of water ice is trapped underground. One region in the northern plains may contain as much water as Lake Superior, the largest of the Great Lakes on Earth.

Other evidence of water on Mars includes bright material that is scattered on the landscape after fresh meteorite impacts. NASA satellites have watched as the exposed material, likely water ice, slowly vaporized in the thin Martian atmosphere. Ice has also been seen sheltered in the shady areas of large craters. It is also possible that seasonal dark streaks appearing on steep slopes may be caused by salty water seeping downward. Or, the streaks could simply be caused by dark sand rolling downhill. More investigation will be needed.

Why is the search for water so important? If life ever existed on Mars, it would probably have needed liquid water to arise. If scientists find pools of water under the Martian surface, perhaps some sort of simple microbial life survives to this day. More importantly for future Martian colonists, finding large amounts of water, either liquid or ice, will be important for growing plants, drinking, and possibly producing hydrogen for rocket fuel.

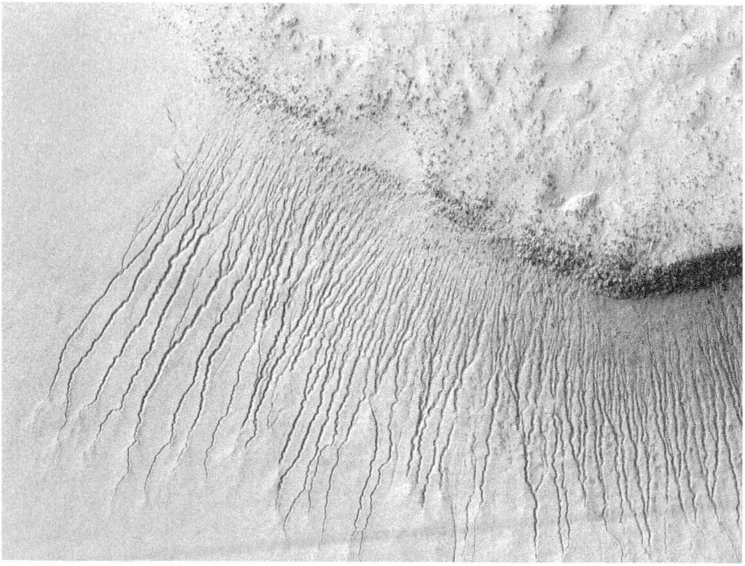

A satellite image of Mars's surface shows gullies and ravines, ranging in width from three to 33 feet (1 to 10 m), flowing from the side of an impact crater. Could they have been caused by water seeping from the sides of the crater?

CHAPTER 23
THE MOONS OF MARS

MARS HAS A PAIR OF MOONS, PHOBOS AND DEIMOS, NAMED AFTER characters in ancient Greek mythology. Phobos means "fear" (which is where we get the word "phobia") and Deimos means "dread." Both characters traveled with their father, Ares, into battle. Ares was the Greek god of war. The Romans called him Mars, the name given to the Red Planet. The moons were discovered by astronomer Asaph Hall in 1877.

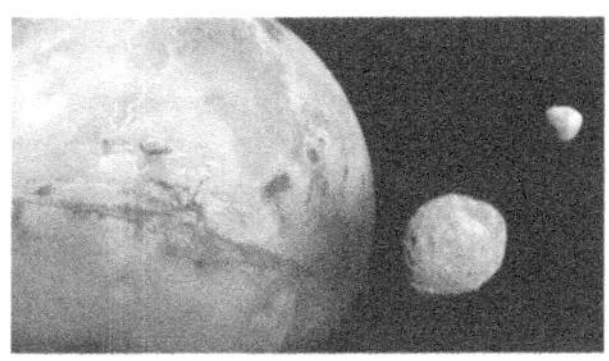

Mars's two moons, Phobos (left) and Deimos.

Both of Mars's moons are tiny, and shaped like lumpy potatoes. The larger of the two moons is Phobos. Its average diameter is about 13.8 miles (22 km). Deimos measures just 7.7 miles (12.4 km). Because they are both so small, astronomers suspect the two moons were once asteroids captured by Mars's gravity.

Deimos rises slowly in the east and sets in the west, just like Earth's Moon. It has many craters, but appears smoother than its twin moon, Phobos. Deimos orbits Mars about every 30 hours.

Phobos is pockmarked by many craters. Stickney Crater is six miles (10 km) in diameter, almost half the diameter of the moon itself! From the surface of Mars, Phobos appears to rise in the west and set in the east. Its orbit is low and fast, circling the planet about twice each Martian day. Mars's gravity pulls Phobos closer and closer. In 30 to 50 million years, the moon will either be pulled apart by Mars's gravity or crash onto the planet's surface.

CHAPTER 24
LIFE ON MARS?

COULD THERE POSSIBLY BE LIFE ON MARS? PEOPLE HAVE BEEN ASKING that question about our close neighbor in the sky for hundreds of years. Four critical elements—carbon, oxygen, nitrogen, and hydrogen—are needed to support life as we know it. Mars has all of these elements today. We also know that in the planet's past, Mars had plenty of water to support whatever life forms may have arisen. Could we somehow detect those life forms, living or long dead? Perhaps there are fossils to be found among the Martian rocks. Now that NASA and other space agencies are sending sophisticated satellites and rovers to Mars, we may someday soon find an answer.

There are certainly no plants or animals on the surface of Mars today. The air is too thin, and temperatures are far too cold. Also, without a thick atmosphere for protection, radiation from the Sun does long-term harm to any complex life form. However, there is a chance that microscopic organisms could still be alive, perhaps left over from the planet's earlier, warmer days. Maybe someday we'll discover them in underground reservoirs of water, or some other sheltered place. Future missions to Mars will almost surely continue looking for signs of life on the Red Planet.

CHAPTER 25
EXPLORING MARS

MARS IS A WORLD THAT HAS CHANGED OFTEN DURING ITS LONG history. It has many features similar to Earth. It has seasons, volcanoes, weather, and canyons. It has polar ice caps, and was once home to oceans of liquid water. Unlike Earth, Mars has an atmosphere that is extremely thin, and it has a much weaker magnetic field. All these traits make Mars a tempting place to study and explore.

Humans first began exploring Mars without ever leaving the safety of Earth. During ancient times, many cultures all over the world were aware of the mysterious Red Planet, including astronomers from Europe, China, India, and the Middle East. When telescopes became strong enough and widely available, some surface features came into view. They included the bright polar ice caps and dark areas that seemed to grow during the planet's summers. Some people speculated these shadowy sections could be forests.

In 1877, Italian astronomer Giovanni Schiaparelli drew the first detailed map of the Red Planet. He noted long, straight lines on the surface, which he called *canali*. The word means "channels" in Italian, but many people thought Schiaparelli was describing canals dug by intelligent Martians. In the 1890s, American

Percival Lowell and other astronomers also claimed to see canals on Mars. These observations were later found to be caused by optical illusions.

By the 1960s, Earth-based telescopes had reached their limit in exploring Mars. If we wanted to learn more, we would have to send spaceships that could observe up close. In July 1965, NASA's Mariner 4 probe reached Mars, beaming back 22 images of the surface of the planet as it zoomed past. The pictures shocked many people. Instead of oceans, forests, and canals, there was desert. Instead of cities, there were thousands of craters. Mars appeared to be a dead place, like the Moon.

In 1971, NASA's Mariner 9 probe arrived, becoming the first spacecraft to orbit an alien world. It captured images that revealed giant volcanoes and a canyon 10 times bigger than the Grand Canyon. Most exciting of all was evidence of water in Mars's distant past.

Later that year, a pair of landers from the Soviet Union touched down on Mars, but they either crashed or failed shortly after landing. Then, in 1975, NASA's Viking 1 and Viking 2 landers safely arrived on the surface of Mars. They chemically analyzed soil samples and sent back high-resolution photos of the Martian landscape.

More missions followed the Viking landers, including high-tech orbiters and rovers. The spacecraft came not only from the United States but also the Soviet Union, the European Space Agency, India, the United Arab Emirates, Japan, China, and Russia. Some of these sophisticated probes were successful, but many were not. Getting to Mars is hard, but scientists learn from their mistakes as well as their successes. We learn more after each mission.

Today, there are several spacecraft orbiting the Red Planet or roving across the surface. NASA's Spirit, Opportunity, and Curiosity rovers lasted many years beyond their mission plans. All of these engineering marvels have given us a greater understanding of Mars. Yet, there are mysteries still to be solved.

Future missions will continue to explore Mars's geology, its atmosphere, and whether it once harbored life. Probes and rovers, however, have their limits. NASA plans to send astronauts to Mars within the next two decades. Whether or not they meet that deadline, humans will someday most likely walk on Mars. They may even live in permanent colonies.

There will be many challenges. Future colonists will need spacesuits and pressurized homes to survive in the cold, oxygen-poor atmosphere. They will need a source of water, and shielding to protect against radiation from the Sun. They will have to grow their own food and make their own fuel and electricity. These are major problems, but scientists believe they can be overcome. Perhaps someday, if people live on Mars long enough, a human baby, the first, true Martian, will be born on the Red Planet.

MARS ORBITERS

CHAPTER 26
JOURNEY TO THE RED PLANET

For thousands of years, people have feared Mars, the Red Planet. They trembled at its rusty color and its strange wanderings through the night sky. Many wondered if an alien civilization thrived on its surface, perhaps plotting an attack on an unsuspecting Earth. After all, in ancient Greek and Roman mythology, Mars was the god of war.

In more modern times, curious astronomers in the late 1800s gazed at Mars through Earthbound telescopes. It appeared similar to Earth, even though its diameter is just 4,220 miles (6,791 km) at the equator, about half the size of our home planet. Even so, Mars is tilted on its axis, just like Earth. That gives the Red Planet seasons. There are polar ice caps and bands of colors on the surface that seem to change from winter to summer. Could that mean, early astronomers speculated, that there could be life on Mars? Scientists were determined to find out.

In the early 1960s, the Soviet Union (most of which was today's Russia) launched a series of probes to explore Mars. All of them failed. In 1964, it was America's turn to try. That year, NASA's Mariner 4 space probe left Earth's orbit, bound for Mars. The discoveries it made changed our way of thinking about the

Red Planet. In more recent years, a fleet of orbiters from several countries has followed Mariner 4, giving us new insight about mysterious Mars.

CHAPTER 27
MARINER 4

- Mission: Mars flyby
- Launch: November 28, 1964
- Launch Vehicle: Atlas-Agena D
- Mars Arrival: July 14, 1965
- Mission End: December 21, 1967
- Spacecraft weight (mass): 575 pounds (261 kg)

Before sending orbiters to circle Mars, NASA scientists first had to prove they could send a spacecraft across vast distances. Mariner 4's mission was to travel past Mars (a flyby) and take pictures to radio back to Earth. It was built tough to survive the harsh, cold conditions of deep space, and used four solar arrays to generate electricity.

Mariner 4 successfully launched at Cape Canaveral, Florida, on November 28, 1964. It rode atop a mighty Atlas-Agena rocket, bound for Mars. (The Mariner 3 Mars mission had failed earlier that month.) During the eight-month voyage, the spacecraft used the Sun and a bright star called Canopus to guide it through space. It traveled farther than any human-made object at that time.

On July 14, Mariner 4 flew about 6,118 miles (9,846 km) over the rusty-colored surface of Mars. The probe's instruments came to life. Its primitive television camera (the best technology available) took 22 low-resolution pictures, covering about one percent of the planet. The images took a week to reach Earth because of slow computer speeds. The pictures revealed Mars was a wasteland, a dry, rocky world filled with Moon-like craters. Mariner 4's other scientific instruments discovered Mars has an atmosphere that is extremely thin. Liquid water would quickly evaporate in the Martian air. These facts shattered the hope that Mars was an inviting place that could harbor life. But the Red Planet still held many surprises, which scientists would soon discover.

The first close-up image of Mars taken by Mariner 4.

CHAPTER 28
MARINERS 6 AND 7

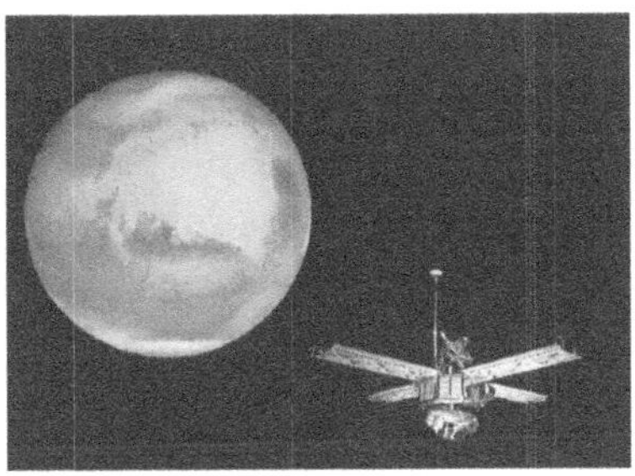

- Mission: Mars flyby
- Launch: Mariner 6: February 24, 1969
- Mariner 7: March 27, 1969
- Launch Vehicle: Atlas-Centaur
- Mars Arrival: Mariner 6: July 30, 1969
- Mariner 7: August 4, 1969
- Mission End: Mariner 6: July 31, 1969
- Mariner 7: August 5, 1969
- Spacecraft weight (mass): 910 pounds (413 kg)

After the success of Mariner 4, NASA prepared more flyby missions. Mariner 5 went to Venus, but Mariners 6 and 7 were sent to Mars. They arrived at the Red Planet within four days of each other in the summer of 1969. Both spacecraft were modified versions of Mariner 4. They were bigger and heavier, and had improved scientific instruments on board. They carried fully automatic instrument packs that included two television cameras. They were also equipped with infrared and ultraviolet spectrometers.

Like Mariner 4, when Mariners 6 and 7 arrived at Mars, they took photographs and instrument readings. They then sped past the planet, slipping into the void of deep space. During their flyby missions, the probes took a combined 210 photos of Mars, increasing the close-up coverage of the planet to nearly 10 percent of its surface.

Mariners 6 and 7 confirmed much of what Mariner 4 had already revealed. The desert-like surface of Mars was dotted with thousands of craters. Also, the Martian atmosphere was very thin, and there was no trace of a magnetic field like that on Earth. Mariner 7 recorded a surface temperature at the south pole of -190 degrees Fahrenheit (-123 degrees C). That made scientists believe that the Martian polar caps were composed of sheets of frozen carbon dioxide.

CHAPTER 29
MARINER 9

- Mission: Mars orbiter
- Launch: May 30, 1971
- Launch Vehicle: Atlas-Centaur
- Mars Arrival: November 14, 1971
- Mission End: October 27, 1972
- Spacecraft weight (mass): 2,200 pounds (998 kg)

In 1969, Mariners 6 and 7 discovered exciting facts about Mars. They also raised more questions. To solve these new riddles, NASA scientists called for something more than a simple flyby mission. They needed a probe that could orbit the planet and investigate it for a long time. Two years later, they got their wish.

Mariner 9 was larger than the earlier Mariner probes, tipping the scales at 1,245 pounds (565 kg). Much of the extra weight came from a larger engine and additional fuel to help Mariner slow down and enter orbit around Mars.

Mariner 9 lifted off from Cape Canaveral on May 30, 1971. (Earlier that year, Mariner 8 failed 20 seconds after launch and crashed into the Atlantic Ocean.) Mariner 9's liftoff went smoothly, beginning a 247 million mile (398 million km) journey through deep space. Five months later, on November 14, 1971, the probe slipped into an elliptical trajectory around Mars, becoming the first spacecraft to orbit another planet. After two rocket burns to correct its orbital path, it was ready to start its mission. At its closest approach, it traveled just 1,025 miles (1,650 km) above the Martian surface, much closer than any of the other Mariner probes.

Mariner 9's mission was delayed by a massive dust storm that obscured most of the planet. Luckily, NASA could reprogram the spacecraft's computer. They shut off the probe's cameras to conserve energy while it waited out the storm.

After several weeks, Mariner 9's instruments were switched back on. The spacecraft began sending high-resolution photos back to Earth, along with other scientific measurements. For almost a full year, it returned 7,329 images of Mars, covering nearly 85 percent of the planet's surface.

Scientists were astonished by the dramatic landscapes they witnessed. A series of enormous shield volcanoes was discovered. The largest was Nix Olympica, which was later renamed Olympus Mons. It towers 16 miles (26 km) above the surrounding plains. The extinct volcano has a diameter of 374 miles (602 km), about the same area as the state of Arizona.

Mariner 9 took photos of Valles Marineris, an immense canyon about 2,485 miles (4,000 km) long. The spacecraft also took the first close-up views of Mars's two rocky moons, Phobos and Deimos.

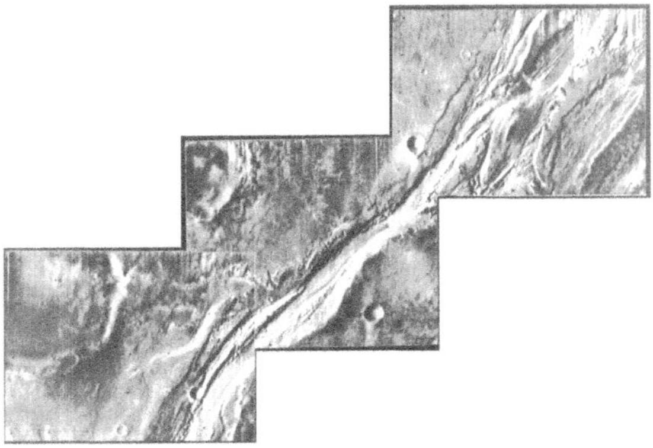

A mosaic of Mariner 9 images shows an eroded valley that may have been formed by flowing water.

Perhaps the most exciting images were those of ancient, dry riverbeds. Scientists became excited by the possibility that, if water once flowed on Mars long ago, microbial life may yet cling to the planet, maybe deep underground.

After a mission lasting nearly one year, Mariner 9 ran out of fuel and could no longer be controlled. The spacecraft was switched off on October 27, 1972. It remains in orbit around Mars today, and will probably crash into the planet sometime in the early 2020s.

CHAPTER 30
MARS 2 AND 3

By 1971, the Soviet Union had launched many Mars-bound spacecraft. The Soviets used powerful rockets to lift heavy, complicated orbiters. Their spacecraft designs were ambitious, but they couldn't quite match the Americans in performance. Several blew up immediately, or were lost shortly after liftoff. Some made it into space but later lost radio contact with Earth. Despite the setbacks, the Space Race between the two countries continued.

In May 1971, the Soviets sent a pair of orbiters named Mars 2 and Mars 3. They were huge, each weighing over five tons (4.5 metric tons). Unlike NASA's Mariner 9, they also included landers to explore the Martian surface.

Mars 2 and Mars 3 arrived at the Red Planet a few days after Mariner 9. They encountered the same planet-wide dust storm that caused the American spacecraft to delay its mission. Unfortunately, the on-board Soviet computers could not be reprogrammed from Earth. The probes circled Mars and dutifully took dozens of photos of the surface-obscuring clouds swirling in the atmosphere.

The Soviet landers also had bad luck. The Mars 2 lander crashed (becoming the first human object to reach Mars). Mars 3's lander touched down safely with the help of parachutes and rockets. It sent a television image for a few seconds, and then contact was lost. Both Soviet missions, so close to turning into exciting breakthroughs, instead were disappointing failures.

CHAPTER 31
VIKING 1 AND 2

- Mission: Mars orbiter/lander
- Launch: Viking 1: August 20, 1975
- Viking 2: September 9, 1975
- Launch Vehicle: Titan III/Centaur
- Mars Arrival: Viking 1: June 19, 1976
- Viking 2: August 7, 1976
- Mission End: Viking 1 orbiter: August 17, 1980
- Viking 2 orbiter: July 25, 1978
- Spacecraft weight (mass), orbiter only, with fuel: 5,125 pounds (2,325 kg)

In 1972, the Mariner 9 orbiter sent back exciting high-resolution images of the surface of Mars. There were valleys and gullies that were apparently formed by erosion, probably from flowing water. Since water is one of the most important ingredients for life as we know it, could creatures have once existed on the Red Planet?

The cold, harsh conditions on Mars today make life nearly impossible. But could there be simple organisms thriving underground? That is a hard question to answer from high above. NASA was determined to take the next step: it would look for life by landing a spacecraft, a miniature science laboratory, on the surface itself.

Viking 1 liftoff.

On August 20, 1975, Viking 1 blasted off from Florida's Cape Canaveral on a pillar of fire and smoke. After a 10-month journey through deep space, it arrived at Mars. It had a twin, Viking 2, which entered orbit a few weeks later.

Each spacecraft included two parts. A lander would conduct experiments on the Martian surface. An orbiter would take high-resolution color images while circling the planet. It would be the most complex mission ever to explore Mars.

The Viking 1 lander safely touched down on Mars on July 20, 1976. The Viking 2 lander followed on September 3, 1976. They took breathtaking photos and conducted many science experiments, including an analysis of soil samples taken from just under the surface. The probes found many clues, but no firm evidence of life.

Meanwhile, high above the planet, the Viking orbiters carried out their missions. They were much bigger than the previous Mariner probes. Each weighed 5,125 pounds (2,325 kg). They were heavier because they carried more instruments and they needed more fuel to slow the spacecraft down when they reached Mars orbit.

The orbiters measured about eight feet (2.4 m) across. They held scientific instruments, cameras, and antennas. Power came from four solar arrays. With the panels spread out, each orbiter spanned about 32 feet (9.7 m) across. Mounted on the side of each spacecraft was the lander, nestled safely inside an "aeroshell." The landers piggybacked to Mars and then dropped to the Martian surface.

Even though the landers were the stars of the mission, the two Viking orbiters had very important jobs. Their high-resolution cameras helped scientists choose the safest sites possible for the landers. They eventually mapped most of the planet's surface, sending about 52,000 color images to Earth by radio, including detailed views of Mars's two moons, Phobos and Deimos. The orbiters also discovered that permafrost, frozen water, lies beneath the Martian surface. In addition, the ice cap at the north pole contains vast amounts of frozen water.

The Viking orbiters lasted far longer than expected. After several years and hundreds of orbits, each spacecraft was eventually powered down by NASA. The Viking missions were a phenomenal success, but much was yet to be learned about mysterious Mars.

More than 100 images, taken by the Viking 1 orbiter, were combined to create this image of Mars. In the middle of the photo is Valles Marineris, a canyon about 2,485 miles (4,000 km) long and six miles (10 km) deep.

CHAPTER 32
MARS GLOBAL SURVEYOR

- Mission: Mars orbiter
- Launch: Nov. 7, 1996
- Launch Vehicle: Delta II
- Mars Arrival: Sept. 12, 1997
- Mission End: Nov. 2, 2006
- Spacecraft weight (mass): 2,272 pounds (1,031 kg)

In August 1993, NASA lost its Mars Observer spacecraft. It was designed to study the Red Planet's climate, magnetic field, and atmosphere. It was also equipped to take high-resolution photos of the surface. Unfortunately, the probe suffered a fatal fuel leak, and all contact was lost just three days before it was due to orbit Mars.

NASA replaced the doomed Mars Observer with a new, lower-cost orbiter called Mars Global Surveyor. On September 11, 1997, it arrived at Mars. Once it reached the planet, it used a novel method to slow down called aerobraking. It went into a long, looping orbit. When it was closest to Mars, the spacecraft dipped slightly into the atmosphere. This caused a force called drag to gradually decrease the orbiter's speed. After repeated orbits lasting over a year, it settled into a circular trajectory just 280 miles (451 km) above the Martian surface.

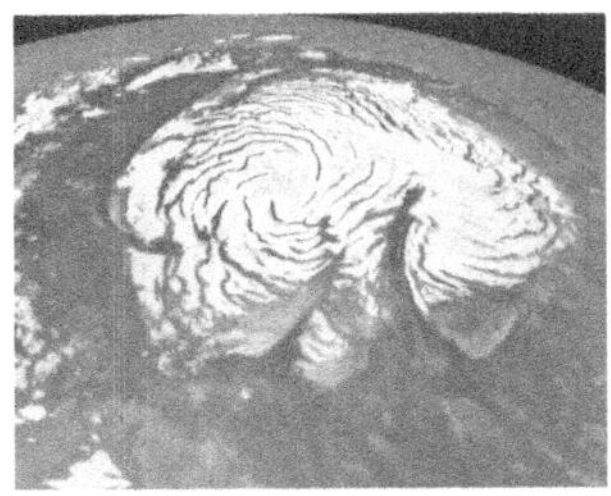

Mars's north pole

Mars Global Surveyor was immensely successful. For over nine years, the sturdy orbiter took thousands of high-quality photographs of Mars and its two moons. Using the probe's data and images, scientists learned about the role of water and how it carved parts of the Martian landscape in the distant past. And it studied Mars's atmosphere and weather patterns. It even captured a real-time image of a dust devil and the tracks left behind in the soil.

CHAPTER 33
2001 MARS ODYSSEY

- Mission: Mars orbiter
- Launch: April 7, 2001
- Launch Vehicle: Delta II
- Mars Arrival: October 24, 2001
- Mission End: Continuing
- Spacecraft weight (mass): 829 pounds (376 kg)

In 1999, NASA lost its Mars Climate Orbiter spacecraft, which was designed to study Mars's atmosphere and climate. That same year, the space agency also lost the Mars Polar Lander, which crashed on the surface. Losing two Mars spacecraft in one year was a disaster. Losing the Mars Climate Orbiter was especially embarrassing. NASA used metric system measurements (kilometers, meters, etc.). Unfortunately, the company that built the spacecraft and supplied its software applied American units (miles, feet, etc.). The small difference in measurement caused the spacecraft to aerobrake too deeply into the atmosphere, where it broke up during its first orbit.

Planetary scientists have a saying that "Mars is hard." Of the many spacecraft that have been sent to the Red Planet, about half have been lost or destroyed for various reasons. Was there a space monster that lurked near Mars, feasting on spacecraft? One NASA engineer joked it was the Great Galactic Ghoul.

In 2001, NASA tried again. On April 7, 2001, the 2001 Mars Odyssey spacecraft blasted off from Florida's Cape Canaveral. The probe was named to honor science fiction author Arthur C. Clarke and his masterpiece, *2001: A Space Odyssey*. After a six-month journey through the void of space, it arrived safely at Mars. As with the earlier Mars Global Surveyor spacecraft, Odyssey used aerobraking to slow down and settle into an orbit around the planet.

The 2001 Mars Odyssey orbiter is shaped like a large box. For power, it uses a three-panel solar array that sticks out of one side. Also projecting from the spacecraft is an antenna, a gamma ray spectrometer, a thermal camera, plus other instruments. They measure the amounts of water, ice, and minerals in the Martian soil.

The spacecraft had a device that measured radiation harmful to astronauts on the planet's surface. (The equipment broke down in 2003, but most of its mission was completed.) Odyssey also acts as a radio relay for Mars rovers that explore the surface.

Odyssey spent years and hundreds of orbits mapping and analyzing Mars. It discovered massive amounts of water ice under the Martian soil. Early in the planet's history, Mars was covered with oceans and rivers. Scientists have always wondered what happened to all the water. Odyssey seems to have solved that mystery: much of the liquid seeped deep into the soil and froze into underground glaciers and polar ice sheets.

The 2001 Mars Odyssey spacecraft continues exploring the Red Planet and acting as a radio relay for surface rovers. As of 2022, it is the longest-serving Mars orbiter. If it stays healthy, it has enough fuel to continue operating at least until the mid-2020s.

This view of Noctis Labyrinthus is a combined collection of images taken by 2001 Mars Odyssey from April 2003 to September 2005.

CHAPTER 34
MARS EXPRESS

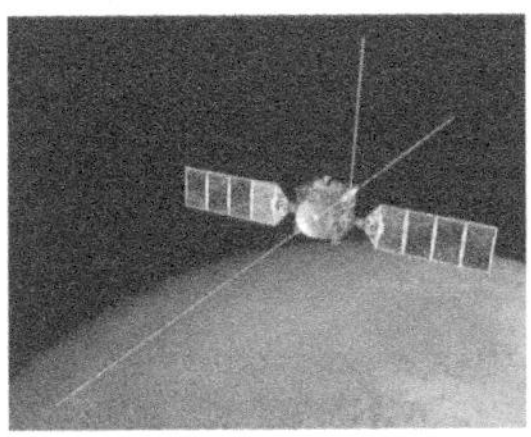

- Mission: Mars orbiter/lander
- Launch: June 2, 2003
- Launch Vehicle: Soyuz FG-Fregat
- Mars Arrival: December 25, 2003
- Mission End: Continuing
- Spacecraft weight (mass): 1,468 pounds (666 kg)

On December 25, 2003, another orbiter arrived at Mars, but it wasn't from NASA. The European Space Agency (ESA) today plays a major role in space exploration. It has been building and flying spacecraft and equipment since the late 1970s. Its members represent 22 European countries, including France, the United Kingdom, Germany, and Italy.

The Mars Express orbiter is ESA's first try at visiting another planet. So far, it has been very successful. As of early 2022, it continues orbiting Mars and sending back data to Earth. It is the second-longest working probe circling Mars (behind NASA's 2001 Mars Odyssey).

The primary goal of Mars Express is to explore the geology and atmosphere of Mars. The probe has a high-resolution stereo camera, as well as sensitive spectrometers and radar. It can detect water and ice as deep as three miles (five km) below the Martian surface.

Mars Express blasted off on June 2, 2003, from the Russian Baikonur Cosmodrome inside the Republic of Kazakhstan. It used Russian-built Soyuz FG-Fregat rockets. It was named Mars Express because of how fast and efficiently it was built. Also, its voyage was fast because Earth and Mars were very close together when the spacecraft was launched.

Several days before Mars Express entered Martian orbit, it released Beagle 2, a lander that was meant to search for signs of life on the surface. Unfortunately, its solar panels did not unfold properly after landing, and Beagle 2 could not communicate with Earth.

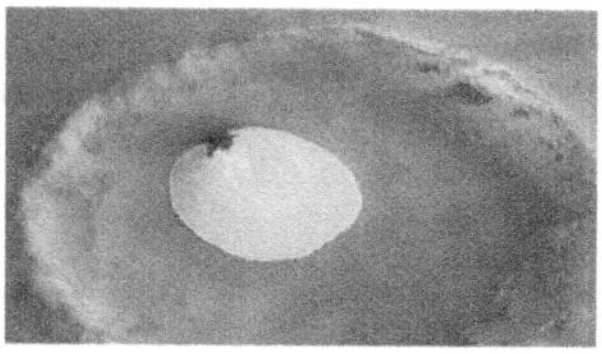

Water ice photographed by Mars Express at the bottom of a crater near the Martian north pole.

The Mars Express orbiter, on the other hand, performed its job with flying colors. It was the first probe to detect water ice at the Martian south pole, hidden underneath thick sheets of frozen carbon dioxide (dry ice). It also detected traces of methane in the atmosphere. The gas may have come from volcanic activity, or it could be evidence of some sort of life form, perhaps microbes in the soil. Future spacecraft will need to investigate further.

Mars Express continues to map the Martian surface and search for water trapped underground. Its cameras have sent home stunning views of the Red Planet. With its large reserves of fuel, the spacecraft should continue operating into the 2020s or beyond.

CHAPTER 35
MARS RECONNAISSANCE ORBITER

- Mission: Mars orbiter
- Launch: August 12, 2005
- Launch Vehicle: Atlas V
- Mars Arrival: March 10, 2006
- Mission End: Continuing
- Spacecraft weight (mass): 2,273 pounds (1,031 kg)

In 2005, NASA readied its next probe to explore the Red Planet. Mars Reconnaissance Orbiter (MRO) had similar instruments as previous spacecraft, including Mars Global Surveyor and 2001 Mars Odyssey. However, MRO included greatly improved technology. Its spectrometers and radar can give the best readings yet of the kinds of minerals found on Mars, and how much water ice is buried beneath the surface.

MRO is equipped with a telescope camera called the High Resolution Imaging Science Experiment (HiRISE). It is much larger and more powerful than any camera previously sent to Mars. It can make out objects as small as three feet (.9 m) across, or about the length of a guitar. (Previous orbiters could pick out surface details about the size of a school bus.)

Technicians installing the Mars Reconnaissance Orbiter High Gain Antenna.

MRO has a huge radio dish antenna, measuring about 10 feet (3 m) across. It was designed to handle the vast amounts of data the orbiter sends back to Earth from its high-resolution camera and scientific instruments. The large antenna and computer capacity are also used to relay many radio signals from other spacecraft, including rovers on the Martian surface.

MRO blasted off from Florida's Cape Canaveral on August 12, 2005. By March 2006, it arrived at Mars. After six months of aerobraking, it settled into a circular orbit about 175 miles (282 km) above the planet.

MRO's high-resolution camera was used to pinpoint safe sites for Mars landers. It can detect medium-sized boulders that might cause a lander to crash. In 2008, NASA's Phoenix Lander spacecraft had its landing site changed after MRO detected too many large rocks in the original target zone. (Phoenix landed safely on May 25, 2008.)

After several years of collecting science data and images, MRO discovered many new things about Mars. It took photos of brand-new craters made by meteorites. Some craters showed water ice scattered in the ground thrown up by the impact. The water ice rapidly evaporated because of the thin Martian atmosphere. The probe took close-up photos of Mars's two moons, Phobos and

Deimos. MRO also captured images of swirling dust devils and found evidence of fresh avalanches in craters and canyons.

By 2015, MRO had sent back to Earth more information than all the previous missions to Mars combined, equaling more than 14 billion pages of printed data. It continues its mission today, and will probably operate until the late 2020s or beyond.

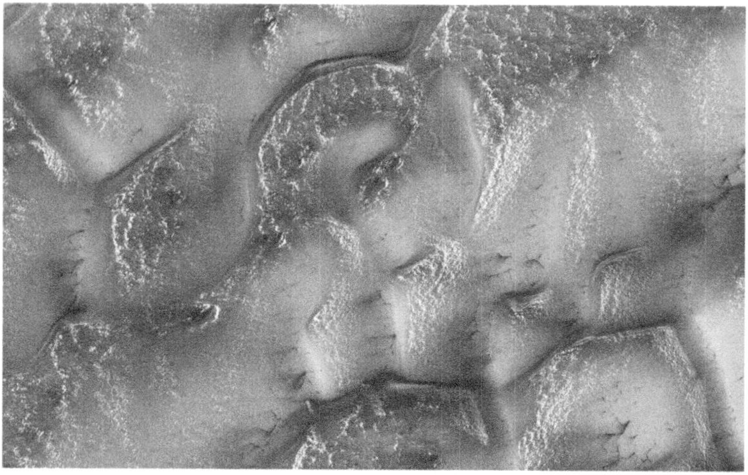

In spring, ice on the smooth surface of some Martian dunes cracks. The escaping gas carries dark sand out from the dune below, often creating beautiful patterns.

CHAPTER 36
MANGALYAAN

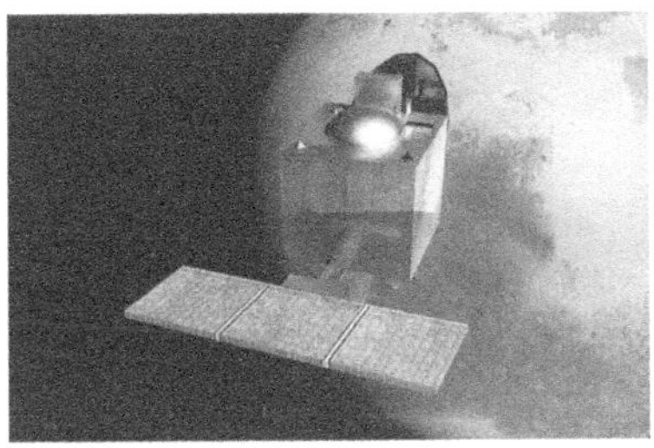

- Mission: Mars orbiter
- Launch: November 5, 2013
- Launch Vehicle: Polar Satellite
- Mars Arrival: September 24, 2014
- Mission End: Continuing
- Spacecraft weight (mass): 2,948 pounds (1,337 kg)

THE MANGALYAAN SPACECRAFT IS A MARS ORBITER FROM INDIA. Also called the Mars Orbiter Mission, it is the country's first interplanetary probe. India's space agency is called the Indian Space Research Organisation (ISRO). It was the fourth space agency to reach Mars, behind NASA, the European Space Agency, and the

Soviet Union's space program. It was the first Asian spacecraft to travel to the Red Planet.

Mangalyaan (a Sanskrit name that means "Mars craft") followed India's successful Moon orbiter in 2008. The Mars mission was planned as a technology demonstration. The scientists at ISRO wanted to prove they could achieve such a goal at a very low cost. Remarkably, they succeeded on their very first try.

Like several other Mars probes, Mangalyaan is shaped like a cube. It has solar arrays projecting from one side. It also has a large antenna and various scientific instruments. They include a spectrometer, a methane sensor, and a high-resolution camera.

Mangalyaan blasted off from India's Satish Dhawan Space Centre on November 5, 2013. It entered Mars orbit on September 24, 2014. It then began exploring Martian surface features, as well as investigating the planet's atmosphere. The spacecraft tracked the amount of dust in the sky, measured ground temperatures, and took rare close-up photos of the backside view of Deimos, one of Mars's two moons. Now that its primary mission is complete, Mangalyaan has enough fuel to continue studying Mars for many years. A more advanced companion orbiter, Mangalyaan-2, is tentatively planned for the mid-2020s.

CHAPTER 37
MAVEN

- Mission: Mars orbiter
- Launch: November 18, 2013
- Launch Vehicle: Atlas V
- Mars Arrival: September 22, 2014
- Mission End: Continuing
- Spacecraft weight (mass): 1,784 pounds (809 kg)

WHY DID MARS DRY OUT? THAT IS THE BIG QUESTION THAT NASA'S MAVEN Mars orbiter was sent to investigate. MAVEN stands for Mars Atmosphere and Volatile Evolution. The spacecraft was designed to study the Martian atmosphere and to find out why so much of it vanished into space.

MAVEN has no camera. Instead, it contains instruments that measure the solar wind and how it erodes Mars's atmosphere. It also has a magnetometer to precisely measure the planet's weak magnetic field.

Without a strong magnetic field (like Earth's), charged particles from the Sun (the solar wind) can cause a planet's atmosphere to slowly be stripped away. Early in its history, Mars had oceans of water and a much thicker atmosphere. But without a protective magnetic field, the solar wind over billions of years reduced Mars's atmosphere by almost 99 percent. Today, the atmosphere is so thin that liquid water cannot exist on the surface because it almost instantly evaporates.

During its primary mission, MAVEN shed light on this process. It also proved that atmospheric erosion greatly increases in intensity during solar storms. NASA plans to continue collecting data from the orbiter into the 2020s.

CHAPTER 38
CUBESATS

InSight is a NASA robotic lander. It was designed to use a seismometer and an underground probe to learn about the deep interior of Mars. InSight launched on May 5, 2018, from California's Vandenberg Air Force Base. On November 26, 2018, it successfully touched down on Mars.

During its long transit, two briefcase-sized spacecraft called CubeSats tagged along with InSight. They shared the same launch vehicle, but separated from InSight during the trip through deep space. The CubeSats were part of a mission called Mars Cube One (MarCO). The twin CubeSats (MarCO A and B) were equipped with radio antennas and solar panels for power.

When the InSight lander descended to the surface, the Cube-

Sats relayed entry, descent, and landing information (EDL) to scientists on Earth in just minutes. Otherwise, communications might have been delayed for hours until other Mars orbiters were in the proper position to relay data. Neither CubeSat went into orbit around Mars. After completing their mission, both spacecraft flew past the Red Planet. They will probably never contact Earth again.

Because they are small, relatively cheap, and efficient, CubeSats could be used by future landers and rovers as inexpensive relays for communicating with Earth, as well as be customized to perform other missions. They can help explore not only Mars, but other planets and objects in our solar system, including the Moon and asteroids.

CHAPTER 39
EXOMARS

The ExoMars Trace Gas Orbiter

ExoMars is a joint mission of the European Space Agency (ESA) and Roscosmos, the Russian space agency. ExoMars comprises four spacecraft on two different rockets launched separately. The first part was launched on March 14, 2016. It reached Mars in October 2016. A lander called Schiaparelli crashed on the surface, but the Trace Gas Orbiter is successfully circling the planet. The second part of ExoMars, which includes a rover, was scheduled to launch sometime in 2022, but the Russian invasion of Ukraine indefinitely grounded the mission.

The ExoMars Trace Gas Orbiter spacecraft is looking for small amounts of methane in the Martian atmosphere. Methane is produced by geological action, such as volcanos. It is also given off by some living creatures, including microbes. The orbiter will also hunt for other trace gases. If it finds methane combined with propane or ethane, that would be a powerful clue that microbes exist on Mars, perhaps under the surface. If the orbiter detects methane with sulfur dioxide, then the methane probably comes from volcanic activity. Besides its search for biological life, the orbiter is also taking high-resolution images of the Martian surface, and can act as a communications relay for future rovers.

CHAPTER 40
JULY 2020 LAUNCH WINDOW AND THE FUTURE

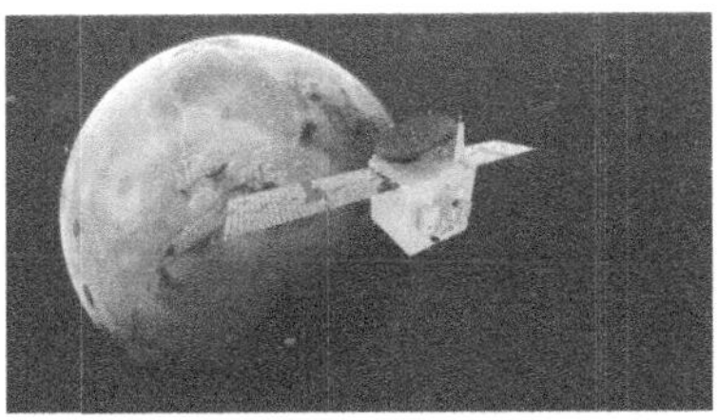

Hope, the Emirates Mars Mission spacecraft.

In July 2020, three Mars-bound spacecraft lifted off from Earth. The countries that took advantage of the narrow launch window included the United States, China, and the United Arab Emirates. The US mission was a direct flight to the Martian surface with the Perseverance rover, but the other two launches included Mars orbiters.

The first of the two orbiter missions to arrive was the Emirates Mars Mission spacecraft, called Hope. It launched on July 19, 2020, and safely achieved orbit on February 9, 2021. About the size of a compact car, it was designed and built by scientists from the United Arab Emirates Space Agency. It is the first Mars mission launched by any Arab or Muslim nation. Its goal is to explore the planet's atmosphere and climate, including dust storms and seasonal weather cycles. The probe's mission duration is expected to last about two years.

In the spring of 2021, Hope's ultraviolet imager captured glowing auroras in the Martian atmosphere. Auroras occur when high-energy particles from a star (such as the Sun) energize atoms in a planet's atmosphere, causing them to shimmer and glow. On Earth, atoms are energized high in the atmosphere in the planet's magnetic field, causing colorful displays we call the northern and southern lights. On Mars, auroras behave differently. Over billions of years, the planet's once-thick atmosphere slowly disappeared, along with its global magnetic field. Today, remnants of the magnetic field are locked into minerals in the Martian crust, creating scattered zones in the atmosphere that interact with charged solar particles. The light shows are similar to the northern and southern lights, but do not exist in the poles like on Earth. Spotting the auroras was the Hope orbiter's first major science finding.

On February 10, 2021, Tianwen-1 arrived in Mars orbit. It was launched on July 23, 2020, by the China National Space Administration. China's first Mars mission, the spacecraft included a lander/rover component, which touched down on Mars on May 14, 2021. High above the surface, the Tianwen-1 ("Questions to Heaven") orbiter relays messages from the rover to controllers on Earth. The orbiter is also packed with its own instruments. It includes a high-resolution color camera, radar to look for subsurface water-ice, a magnetometer to map the planet's magnetic field, and several other scientific instruments. It has a planned mission length of two years.

Space agencies from around the world are designing and constructing future missions to Mars. Technology continues to improve, and hard lessons are learned from past mistakes. New spacecraft sent to Mars build upon the successes—and failures—of the missions that went before them. Our knowledge grows with each journey to the mysterious Red Planet.

MARS LANDERS

CHAPTER 41
LANDING ON MARS

Starting in the mid-1970s and continuing to the present day, spacecraft from Earth have been landing on the mysterious planet Mars. They have included the groundbreaking US Viking landers of the 1970s and the more recent Tianwen-1 lander from China.

To arrive safely on the planet's surface, landers must overcome great odds. They travel millions of miles through deep space, reaching Mars with pinpoint accuracy. Landers drop through Mars's thin atmosphere at hyper speeds that can cause delicate electronics to burn up or shake apart. They slow down with perfectly timed rockets and parachutes to avoid crashing on the rocky ground.

These engineering marvels must land automatically. Mars is so far from Earth that human controllers can't redirect them if something goes wrong. Radio signals take too long to communicate with spacecraft in real-time.

Landers are sent on this perilous journey so we can unlock the secrets of Mars's harsh landscape. The mysteries we solve on Mars may one day help us understand Earth's own history and future.

CHAPTER 42
WHY GO TO MARS?

BEFORE SCIENTISTS KNEW HOW TO BUILD LANDERS THAT COULD SET down on Mars, orbiters blazed a trail to the Red Planet. In late 1971, NASA's Mariner 9 began a nearly yearlong investigation of Mars. The 1,245-pound (565-kg) probe was the first spacecraft to orbit another world besides Earth. At the closest point in its orbit, it came within 1,025 miles (1,650 km) of the surface.

Mariner 9 took thousands of photos and beamed them back to Earth by radio. Analysts who studied the 7,329 images were astonished. Earlier Mariner probes had hinted at a dead, crater-filled planet. But Mars held many surprises.

The photos from Mariner 9 revealed that Mars has huge volcanoes rising above its northern plains. The biggest, Olympus Mons, is 100 times the size of Hawaii's Mauna Kea. It is by far the largest volcano in the solar system. Mars also has a canyon nearly 2,485 miles (4,000 km) long and several miles deep, dwarfing Arizona's Grand Canyon. (The Martian canyon was named Valles Marineris in honor of Mariner 9.)

Most exciting of all was the hint of water on the Martian surface. Pictures from Mariner 9 clearly showed ancient, dry riverbeds that must have flowed with liquid long ago. Scientists also began to suspect that the planet's north polar ice cap contained not just frozen carbon dioxide (dry ice) as first thought, but also huge amounts of frozen water.

The chance that Mars once had vast amounts of water was a thrilling idea. Since water is one of the key ingredients of life as we know it, did life also evolve on Mars? And if water could be found on Mars, maybe buried underground, could life still thrived on the Red Planet today?

In 1971, there was no way to know for sure if Mars harbored life by looking down from an orbiter hundreds of miles above. Science experiments had to be conducted on the surface. Even as Mariner 9 continued to map Mars, scientists in the United States began designing a new kind of spacecraft that could land safely on the planet. These probes are called landers. In the coming years, they would tell us more about the enigmatic Red Planet than ever before.

CHAPTER 43
MARS 2 AND 3

STARTING IN THE MID-1950S, THE UNITED STATES AND THE FORMER Soviet Union (much of which is today's Russia) were locked in the Space Race. They competed to be the first in many areas of space exploration.

In 1970, the Soviet Union's Venera 7 touched down on the planet Venus. It was the first spacecraft to land successfully on an alien world. Unfortunately, it landed too hard.

Damage from the landing and the planet's high temperatures caused the probe to fail. It sent radio signals back to Earth for less than an hour.

The Soviet Union's Mars program was just as ambitious. Sputnik 24 was built to land on the Martian surface. It launched in 1962, but failed to leave Earth's orbit. Then, at the end of 1971, a pair of Soviet spacecraft arrived safely at Mars. They were called Mars 2 and Mars 3. Unlike America's Mariner 9, they included landers. They were big machines, each weighing about 2,700 pounds (1,225 kg) with fuel.

After entering the Martian atmosphere, the spherically shaped landers were designed to fire braking rockets and use parachutes to slow down. Once safely on the surface, four pedals would unfold to make sure the landers were upright. Next, science experiments would be performed on the Martian soil and weather, and the collected data would be radioed to Earth.

The landers also included shoebox-sized rovers that could explore the Martian surface. They used metal skis that lifted and pushed them across the ground, and they could also detect and avoid obstacles.

Mars 2 and Mars 3 were very advanced machines for their time. However, limited technology and faulty design left no room for error. Although the Soviet orbiters had some success, bad luck struck the landers. Both were doomed.

On November 27, 1971, the Mars 2 lander plunged toward the planet at high speed. Its parachute failed to open, and Mars 2 was destroyed on impact. Although it crashed, it became the first human object to reach Mars.

Unlike its twin, Mars 3 landed safely on December 2, 1971. For several seconds, it sent a partial television image of its surroundings. Then, all contact was lost. A planet-wide dust storm may have harmed the probe's delicate electronics, but we'll probably never know for sure what happened.

Both Soviet landers could have made breakthroughs in science. Instead, the world would have to wait for America's turn. Five years later, the history-making Viking landers would be ready to explore Mars.

CHAPTER 44

THE VIKING LANDERS

- Mission: Mars orbiter / lander
- Launch: Viking 1: August 20, 1975
- Viking 2: September 9, 1975
- Launch Vehicle: Titan III / Centaur
- Mars Arrival: Viking 1: June 19, 1976
- Viking 2: August 7, 1976
- Touchdown: Viking 1 lander: July 20, 1976
- Viking 2 lander: Sept. 3, 1976
- Mission End: Viking 1 lander: Nov. 13, 1982
- Viking 2 lander: April 11, 1980
- Spacecraft weight (mass), lander only: 1,262 pounds (572 kg)

In March 1974, two more landers from the Soviet Union tried to beat the United States to Mars. However, bad luck continued to plague the Soviet space program. The Mars 6 lander crashed on the surface, while the Mars 7 spacecraft missed the planet completely. The stage was set for the Americans.

In 1976, twin spacecraft entered orbit around Mars. They were the Viking 1 and Viking 2 probes, which NASA had launched in late 1975. After nearly a year of traveling through deep space, each Viking reached the Red Planet with pinpoint accuracy.

Each spacecraft included two parts. The orbiters were designed to take high-resolution color images from above the planet. The stars of the mission, however, were the landers. They would conduct experiments on the Martian soil, including a search for signs of life. Viking 1 and Viking 2 were the most ambitious spacecraft ever to explore Mars at that time.

In 1971, the NASA Mariner 9 orbiter proved that water flowed on Mars eons ago. Water is a necessary ingredient for life as we know it. Could living creatures have sprung from the Red Planet, as they had on Earth? The scientists who built the Viking landers were determined to find out.

The Vikings were marvels of engineering for their time. About the size of a small SUV, they were made of aluminum and titanium alloys. Each weighed about 1,262 pounds (572 kg). They included scientific instruments, communications gear, and cameras. Landing rockets were mounted underneath.

Each lander contained a miniature biological laboratory to search for microscopic Martian life. Also on board were instruments to monitor the weather, plus a seismometer to detect possible marsquakes. Controlling all the equipment were twin onboard computers.

The landers were sterilized before launching. The search for life would be ruined if the landers inadvertently detected Earth-born bacteria that had mistakenly hitchhiked a ride. To prevent this, the Viking landers were baked at 234 degrees Fahrenheit (112° C) for 40 hours. They were then sealed in a protective shell to keep them germ free. Next, the spacecraft were mounted atop rockets, ready for liftoff.

Viking 2 lifts off.

On August 20, 1975, a powerful Titan III/Centaur rocket blasted off from Florida's Cape Canaveral. The launch went smoothly, sending the Viking 1 orbiter/lander on its long journey to Mars. Less than three weeks later, on September 9, 1975, Viking 2 followed its twin toward the Red Planet. The landers piggybacked on the orbiters all the way to Mars. Once they reached orbit, they separated and began their plunge to the Martian surface.

One of the biggest challenges in designing an unmanned lander is controlling it. In 1969, the Apollo 11 lunar module *Eagle* nearly crashed into a boulder field during its descent to the Moon. Astronauts Neil Armstrong and Edwin "Buzz" Aldrin took control of the spacecraft and steered it away from danger. With Martian probes, there are no astronauts to take over if something goes wrong.

Another problem with interplanetary probes is the vast distance of space. At their closest, Mars and Earth are on average about 35 million miles (56 million km) apart. The Viking radio signals, traveling at the speed of light, would take about three minutes to reach Earth. It would then take another three minutes for NASA to send a signal back to Mars. (By comparison, radio signals from the Moon take just one second.) Earth and Mars would be even farther apart when the Vikings were scheduled to land. If something went wrong on Mars, the crisis would be over long before engineers at NASA could do anything about it. Each Viking lander had to carry out its landing with no help from controllers on Earth.

Another difficulty with landing a spacecraft is decelerating enough to avoid a catastrophic impact on the surface. At first, each Viking craft would use Mars's own atmosphere to slow down. Even though the air on Mars is very thin, spacecraft can employ a technique called "aerobraking." This uses the friction of moving through the air at high speed to reduce velocity, but the process generates tremendously high temperatures. To avert damage, the landers and their delicate electronics were protected by saucer-shaped heat shields.

CHAPTER 45
TOUCHDOWN!

On July 20, 1976, the Viking 1 lander separated from the orbiter and began its descent. (By coincidence, it was the seven-year anniversary of the Apollo 11 Moon landing.) Cameras aboard the Viking orbiters helped analysts pick the safest place to land, in a part of Mars called Chryse Planitia. Anxious scientists on Earth held their breath as the lander began its long, arching fall to the planet's surface.

The lander sped through the upper atmosphere at over 10,000 miles per hour (16,093 kph), about four times the speed of a bullet. Aerobraking slowed the spacecraft down. Friction caused the outer surface of the heat shield to reach a scorching 2,700 degrees Fahrenheit (1,482° C), but the lander was protected inside.

By the time Viking 1 was 3.7 miles (6 km) above the surface, it had slowed down to about 600 miles per hour (966 kph). A billowing, 52-foot (16-m) cloth parachute was shot out of a small cannon. Explosive bolts jettisoned what was left of the charred heat shield. The spacecraft's three landing legs unfolded, spring-loaded to cushion the impact with the ground.

The massive parachute helped Viking 1 decelerate to just over 134 miles per hour (216 kph). That was still too fast. At an altitude of about .9 mile (1.5 km), the parachute detached. Three radar-controlled retrorockets ignited under the lander, slowing it even more.

At 8:12 a.m. Eastern Daylight Time, the Viking 1 lander gently touched down on Mars after a journey of 440 million miles (708 million km). It impacted the surface at about 5.4 miles per hour (8.7 kph), about the same force as jumping off a tall kitchen stool.

Viking 1's engines shut down, and silence returned to the Martian surface. For 19 long minutes, NASA controllers on Earth could only sit and wait for a signal. Was the lander safe, or had it crashed? Then word raced through Mission Control: "Touchdown! We have touchdown!"

Shortly after landing, Viking 1 activated its twin cameras. The first image sent back to Earth was a black-and-white picture showing one of Viking 1's landing pads resting in the rocky soil. Later, color images revealed a flat, reddish Martian landscape with a salmon-pink sky (caused by small dust particles in the air).

The first clear photo from the surface of Mars was taken just minutes after the Viking 1 lander set down. One of the lander's footpads is seen in the lower right corner. NASA scientists took this photo first so they could see how the footpad affected the Martian soil.

Less than seven weeks after Viking 1's success, it was joined by Viking 2. On September 3, 1976, the lander touched down safely at Mars's Utopia Planitia, about 4,000 miles (6,437 km) west of Viking 1.

The surface of each landing site was strewn with rocks and boulders of various shapes and sizes. The landscape resembled the deserts of the American Southwest. An orange-colored, sandy soil covered the ground. It lay in drifts, thanks to the Martian breezes.

Later, the Viking landers chemically analyzed the soil. It was like a loosely packed iron-rich clay. It was reddish because of a chemical reaction with the iron called oxidation—rust.

After a full day on the planet, Viking 1 issued the first weather report from Mars: "Light winds from the east in the late afternoon... Maximum wind, 15 miles per hour (24 kph). Temperature ranging from -122 degrees Fahrenheit (-86° C) just after dawn to -22 degrees Fahrenheit (-30° C)."

The Viking landers took the first close-up photographs of Mars. The high-resolution images were in black-and-white and in color. In total, the landers sent back over 4,500 images of the Martian surface, delighting scientists, and everyone else, on Earth.

Each Viking lander included three biology laboratories. A six-foot (1.8-m) -long robotic arm scooped up soil samples and delivered them to the labs. Several tests were designed to look for signs of life. For example, Earth plants "eat" carbon dioxide gas, and this can be measured. Other tests looked for tiny organisms in the Martian soil.

At first, the Viking tests for life were promising. But when the data was examined further, most scientists believed that no Martian life was detected. Many experts think the only way to tell for sure is to bring a soil sample back to Earth for further study—or send human astronauts to Mars.

The Viking landers were meant to last about 90 days on the surface of Mars. They actually lasted for several years, working far beyond their mission. The Viking 2 lander finally stopped sending signals to Earth on April 11, 1980. The Viking 1 lander ceased operation on November 13, 1982.

Much of the technology used on later Mars missions was pioneered by the Viking landers. Many questions remain, but the science they performed was a giant step forward in learning about the Red Planet.

Viking 2 takes a "selfie" with the rocky field of Mars's Utopia Planitia behind it. The lander's American flag decal is seen clearly in the image.

CHAPTER 46
MARS PATHFINDER

- Mission: Mars lander / rover
- Launch: December 4, 1996
- Launch Vehicle: Delta II
- Mars Touchdown: July 4, 1997
- Mission End: September 27, 1997
- Spacecraft weight (mass): 1,020 pounds (463 kg)

After the successful American Viking landers, the Soviet Union continued its own efforts to send spacecraft to Mars. In 1988, they launched Phobos 1 and Phobos 2. The probes were sent to study Mars and its two moons, Phobos and Deimos. They were also designed to drop landers on Phobos to examine the moon up close. Unfortunately, Phobos 1's computer shut down the probe's systems while it was between Earth and Mars. Phobos 2 suffered a similar fate. Its computer failed just before the lander was ready to be dropped onto the Martian moon.

In 1991, most of the Soviet Union became the Russian Federation. The Russian space program tried again to reach Mars in 1996. Unfortunately, they were met with more failure. The Mars 96 orbiter/lander spacecraft crashed into the Pacific Ocean shortly after launch.

In 1997, the United States finally returned to Mars. Over 20 years had passed since the Viking missions. This time, NASA sent a spacecraft called Mars Pathfinder. Piggybacking on the lander was a microwave-oven sized rover called Sojourner. (It was named after US Civil War abolitionist Sojourner Truth. The word sojourner means "traveler.")

Pathfinder was part of NASA's Discovery Program. Its motto was "better, faster, cheaper." Each mission was designed to be built quickly and for relatively little money (for a spacecraft). Pathfinder's simple mission: test the Sojourner rover in the hostile environment of Mars. The Pathfinder lander itself would take pictures and perform science experiments.

Mars Pathfinder lifted off from Florida's Cape Canaveral on December 4, 1996. It traced a long, arching path through space. By the time it caught up with Mars in July 1997, it had traveled over 300 million miles (483 million km).

On July 4, 1997, Pathfinder landed on the rocky Ares Vallis plain of Mars. A large parachute helped slow it down through the atmosphere. A unique system of oversized airbags cushioned the final impact. Pathfinder survived the landing in great shape and soon began its work.

Even though the Sojourner rover was the star of the mission, the Pathfinder lander flawlessly performed its own science experiments. It monitored the Martian weather, including temperature, wind speed, and atmospheric pressure. The first night on Ares Vallis, the temperature dropped to a bone-chilling -64 degrees Fahrenheit (-53° C).

The Mars Pathfinder lander was also equipped with a stereoscopic camera. It was mounted on a pop-up mast that boosted the camera several feet above the ground. It took more than 16,500 high-resolution images of the surrounding Martian landscape in black-and-white, color, and 3-D.

Pathfinder lasted many weeks beyond its expected lifespan of 30 days. Finally, on September 27, 1997, the lander lost contact with Earth after nearly three months. The harsh conditions on Mars, especially the cold, probably caused its computer or batteries to fail.

The Pathfinder lander was officially renamed the Carl Sagan Memorial Station. Carl Sagan was a popular astronomer who had died earlier in 1997. His work greatly helped in the exploration of Mars.

A panorama view of Mars taken by Pathfinder. The Sojourner rover can be seen inspecting a large boulder on the right.

CHAPTER 47
MARS POLAR LANDER

Scientists who send spaceships to Mars work knowing that it is a monumentally difficult task. So many things can go wrong, from launch explosions to crashing while trying to land, and everything in between. The success of the Mars Pathfinder mission gave NASA managers confidence that future Mars missions really could be achieved quickly and cheaply.

The Mars Polar Lander was a 640-pound (290-kg) spacecraft. Its primary mission was to search for water ice near the Martian south pole. It carried two robotic bullets that would be fired into the ground just before the probe landed. They would burrow deeply into the soil, where the harmful effect of the Sun's rays had not altered the dirt's chemistry. That information could tell scientists much about the early geology of Mars.

The lander was launched on January 3, 1999. It reached Mars on December 3, 1999. It was aiming for a region called Planum Australe. Everything seemed to be going well as the spacecraft entered the planet's thin atmosphere and streaked toward the ground. Suddenly, contact with the spacecraft was lost. Many minutes passed with no signal. Finally, Mars Polar Lander was declared lost.

A team of researchers later discovered that as the probe

approached the surface, its landing legs extended as planned. However, the vibrations from the leg extension caused the on-board computer to think it had already landed, so it shut down the retrorockets too soon. The lander went into free fall the last 100 feet (30 m) to the frozen ground. Mars Polar Lander was demolished and the mission proved to be an embarrassing failure. Mars was still hard, after all.

CHAPTER 48
BEAGLE 2

THE NEXT LANDER TO EXPLORE MARS DID NOT COME FROM THE United States or Russia. Instead, it came from Europe. The European Space Agency (ESA) has been building and flying spacecraft and equipment since the late 1970s. Its members represent 22 countries, including France, the United Kingdom, Germany, and Italy.

The Beagle 2 lander was built by scientists from the United Kingdom. Its mission was to search for signs of life in the Martian soil. It had tools that could scoop up and analyze dirt and rocks with a robotic arm.

Beagle 2 piggybacked aboard the Mars Express orbiter. It was the ESA's first try at visiting another planet. Mars Express launched on June 2, 2003. Although the orbiter was very successful, Beagle 2 failed soon after landing.

Shaped like a large bowl, Beagle 2 weighed about 73 pounds (33 kg) and measured 3.3 feet (1 m) in diameter. It had four solar panels it used for power. A few days before Mars Express reached its destination, the lander detached and followed its own course. On December 25, 2003, it began its descent to the Martian surface. At precisely the right altitude, parachutes slowed its speed, and airbags cushioned its impact with the ground. However, attempts to contact the spacecraft after landing were met with silence. The mission had failed.

Later, images from NASA orbiters revealed that Beagle 2's solar panels had not unfolded properly after touching down. The radio transmitter was blocked, and it could not produce the power it needed. Beagle 2 could not call home for help, and it soon succumbed to its sad fate.

CHAPTER 49
PHOENIX MARS LANDER

- Mission: Mars lander
- Launch: August 4, 2007
- Launch Vehicle: Delta II
- Mars Touchdown: May 25, 2008
- Mission End: November 2, 2008
- Spacecraft weight (mass): 770 pounds (349 kg)

In 2002, NASA's Mars Odyssey orbiter discovered that the north polar region of Mars had huge amounts of water ice buried under the surface. NASA decided to try a second time to land a spacecraft near a Martian pole. Mars Polar Lander had failed in 1999 at the south pole. This time, a NASA spacecraft would land on the opposite side of the planet. Its name was Phoenix Mars Lander. (Phoenix is the bird of ancient myths that was reborn from fire.) The new spacecraft's design would be based on the earlier lander, but with many improvements.

Phoenix had two mission goals. The first was to study the history of Martian water. To do this, it would analyze samples of the arctic ice. Water is a necessary ingredient of life. The lander's second goal was to examine whether the arctic soil on Mars could support life. By digging deep into the ground, perhaps it detect bacteria lying dormant in the dry, bitterly cold Martian environment.

Phoenix was equipped with a long robotic arm, designed to dig deeply into the soil and bring samples to be studied by instruments on the lander. A camera was mounted on the scoop end of the arm. Phoenix was also furnished with microscopes, gas analyzers, devices to study the Martian weather, and a high-resolution stereo camera mounted on a tall mast rising above the lander.

Phoenix Mars Lander blasted off from Florida's Cape Canaveral on August 4, 2007. Its journey to Mars went smoothly, and it arrived on May 25, 2008. As the probe descended to the surface, it used a parachute to slow its speed. From high above, NASA's Mars Reconnaissance Orbiter captured an image of Phoenix with the parachute deployed.

During the last landing stage, Phoenix released its parachute and fired retrorockets. It touched down safely near the north pole in the Vastitas Borealis region, the site of a large ocean billions of years ago. Phoenix became the first lander to successfully land in a polar region of Mars.

Photos from Phoenix soon arrived at Mission Control on Earth. Unlike other landing sites, the images revealed a very flat plain with few large rocks. Directly underneath the retrorockets were white patches. These were areas swept clean of loose soil by the rockets, revealing hard layers of water ice. For the first time, a lander had made direct contact with water on Mars!

Phoenix conducted science experiments for several months. It sent weather reports, dug for soil samples, and tested water ice.

No microorganisms were found, but analysis of the ice revealed that flowing water was present near the poles in Mars's distant past.

Phoenix continued operating beyond its 90-day lifespan, but not for long. After 155 days on Mars, the lander fell silent, the victim of the coming winter at the north polar region, where temperatures had been falling to -142 degrees Fahrenheit (-97° C). Finally, the cold drained the lander's batteries faster than they could be recharged by the solar panels. NASA declared the mission over on November 2, 2008.

Phoenix uncovered water ice using its robotic arm.

CHAPTER 50
SCHIAPARELLI EDM LANDER

In the early 2010s, the European Space Agency (ESA) teamed up with Roscosmos, the Russian space agency. They planned twin missions called ExoMars to look for past life on the Red Planet.

The first ExoMars mission contained two spacecraft. The ExoMars Trace Gas Orbiter was designed to look for small amounts of methane and other gasses in the Martian atmosphere. Scientists hoped to detect the by-product of life on the surface.

The other part of the mission was the Schiaparelli EDM lander. EDM stands for Entry, descent and landing Demonstrator Module. The spacecraft was named after the famous Italian astronomer Giovanni Schiaparelli. Schiaparelli EDM's mission was to test technology for future landers and rovers. It would also measure weather conditions, including electricity in the atmosphere. The spacecraft launched on March 14, 2016, and reached Mars in October 2016. (The second part of ExoMars, which includes a rover, is scheduled as of this writing to arrive at Mars early in the summer of 2023.)

The ExoMars Trace Gas Orbiter successfully went into orbit around Mars and commenced its mission. Schiaparelli EDM, however, was not so lucky. On October 19, 2016, it entered the Martian atmosphere at 13,000 miles per hour (20,921 kph). It then began its descent to the Meridiani Planum region. A parachute and retrorockets helped slow it down. But instead of landing softly, the retrorockets shut down too soon. The spacecraft smashed into the ground at about 186 miles per hour (300 kph) and was obliterated.

Images later taken by NASA's Mars Reconnaissance Orbiter show the black scorch mark on the surface where Schiaparelli EDM crashed. Although the lander was lost, the mission was still a partial success. Unlike the earlier Beagle 2 lander, Schiaparelli EDM transmitted data to Earth as it fell. The landing system had been tested, and lessons were learned for future missions.

CHAPTER 51
INSIGHT

- Mission: Mars lander
- Launch: May 5, 2018
- Launch Vehicle: Atlas V-401
- Mars Touchdown: Nov 26, 2018
- Mission End: Dec 2022 (estimate)
- Spacecraft weight (mass): 1,590 pounds (721 kg)

One of the last unexplored parts of Mars is the planet's deep interior. What kinds of rocks are hidden far below the surface? How big is the core, and is it solid or molten? NASA's InSight lander, with scientific hardware built by European space partners, was built to help scientists answer these questions. By studying Mars's crust, mantle, and core, we can learn much about how the inner solar system's rocky worlds, including Earth, were formed long ago.

InSight (Interior Exploration using Seismic Investigations, Geodesy and Heat Transport) was launched on May 5, 2018, from California's Vandenberg Air Force Base. It arrived at the Red Planet on November 26, 2018, after a trip lasting nearly seven months and spanning 301 million miles (484 million km). The landing zone was the Elysium Planitia region, just south of an area containing major volcanoes.

A large parachute and retrorockets slowed InSight's descent when it entered Mars's atmosphere. After landing softly with its three shock-absorbing legs, two solar panels unfolded. They are designed to recharge the lander's batteries each day. With the panels extended, InSight is about the size of a large automobile. The lander weighs just under 800 pounds (363 kg).

InSight's three-legged design is based on NASA's Phoenix Mars Lander spacecraft, with additional science equipment attached. The most important tools are for studying the planet's interior. The first is a seismometer that measures marsquakes. It is so sensitive it can detect tiny vibrations caused by faraway meteorite strikes. Seismic waves change as they travel through Mars's crust and deep interior. This gives scientists a way to study the internal structure of the planet, including its mantle and core.

InSight's second important instrument is called the Heat Flow and Physical Properties Package (HP^3). It measures heat that rises from the planet's core. An 18-inch (46-cm) spike, nicknamed "the mole," is designed to burrow deep into the soil using an internal hammering device to drive itself about 16 feet (5 m) under the surface. A cable with heat sensors along its length trails the spike into the hole. Scientists can measure the rate at which heat from the planet's interior reaches the sensors. From this data, they can learn about the structure of Mars's core.

Within a few weeks after landing, both the seismometer and the mole were carefully placed on the Martian surface by InSight's 7.8-foot (2.4-m) robotic arm, which has cameras on its end to guide operators on Earth. Besides its primary tasks, the lander is also equipped to measure wind, temperature, and the local magnetic field.

On April 6, 2019, after 128 Martian days (sols) on the planet, InSight's seismometer detected a faint trembling that may be the lander's first recorded marsquake. Scientists are still studying the data from this quake, and other subsequently detected temblors, but the readings mean that Mars is certainly seismically active.

About 90 percent of the quakes detected on Earth are caused by the movement of our planet's tectonic plates. These are giant slabs of rigid crust that slowly move atop a semisolid layer beneath them. As the active tectonic plates move, grind, and slip against each other, they produce tremors that are detected by seismometers.

Unlike Earth, Mars's crust is a single, thick piece, so any marsquakes detected must be caused by something other than plate tectonics. Many scientists believe Mars has a gradually cooling interior. Millions of years ago, it was hot enough to give rise to colossal volcanoes such as Olympus Mons. Today, the ground near these volcanic hotspots is settling, which sends detectable shock waves through the crust.

After a promising start, InSight's self-hammering heat probe known as the mole halted progress after burrowing only about 12 inches (30 cm) deep, at least 9 feet (2.7 m) short of its goal. At first, NASA scientists feared the mole may have been blocked by a large rock. However, they soon realized the problem was the Martian soil itself. The landing site apparently has a different kind of soil than what was encountered by other landers and rovers elsewhere on Mars. The soil near InSight is more cohesive, with particles that stick together. That means that when the mole drills down, the soil clumps and sticks to the walls of the tunnel instead of falling back down into the hole. The spike-like mole depends on the friction created by soil falling back into the hole to continue burrowing. Without enough friction, the mole can't hammer itself any deeper; it merely lurches up and down in the shallow hole, going nowhere.

NASA tried helping the probe by pressing down on the soil next to the hole with a scoop on the end of the lander's robotic arm. Scientists hoped this would help collapse the pit and give the mole the friction it needed to reach a sufficient depth. This process succeeded in driving the back end of the mole about 1.2 inches (3 cm) below the surface, but then all progress stopped. After hundreds of additional hammer strokes, NASA called a stop to the experiment. Tilman Spohn, one of the lead scientists who worked on the device, said in a statement, "We've given it everything we've got, but Mars and our heroic mole remain incompatible. Fortunately, we've learned a lot that will benefit future missions that attempt to dig into the subsurface."

Despite the setback, InSight continued investigating the Red Planet with its remaining instruments. When its primary science goals were accomplished about two years after landing, NASA greenlit an extended mission lasting well into 2022, even though dust that had accumulated on the lander's solar panels caused them to produce less power.

By summer 2022, InSight's seismometer had detected over 1,300 marsquakes. On May 4, 2022, after nearly three-and-a-half years of measuring planetary vibrations, the seismometer registered an estimated magnitude 5.0 quake, almost 10 times as powerful as any temblor yet detected on Mars.

While the seismometer dutifully collected data on marsquakes, the lander's radio Doppler instrument helped scientists measure the thickness of the planet's crust and mantle, as well as the radius of its core. (The crust was thinner than expected, with two to three sublayers.) In addition, the lander's twin set of weather sensors collected data about the wind and temperature at the landing site.

In early summer 2022, NASA announced that InSight's solar panels had become so covered with dense, rust-colored Martian dust that not enough power was being generated to continue the lander's mission. The panels were designed to last two years, which means the lander has had a bonus period of over a year and a half of performing science experiments. As of July 2022, all instruments except the seismometer have been shut down. Scientists expect the lander's batteries to be fully discharged by the end of 2022, rendering the lander inoperable. A passing dust devil, common on Mars, may whirl away some of the dust on the solar panels, but chances of that happening are slim.

In the meantime, as InSight nears of end of its mission, scientists continue to learn about Mars's deep interior, and how it formed. This data will also shed light on how all rocky planets are created, including our home world of Earth.

For updates and more information, visit NASA's InSight web page at: https://www.nasa.gov/mission_pages/insight/main/index.html

MARS ROVERS

CHAPTER 52
JOURNEY ACROSS AN ALIEN WORLD

NASA's Curiosity rover.

HUMANS FIRST LANDED ROBOTIC spacecraft on the surface of Mars during the 1970s. Scientists gathered a huge amount of data from these stationary machines. However, to learn more about the Red Planet, a new generation of science laboratories was launched. They were packed with experiments to test Mars's atmosphere and soil. They bristled with cameras to capture images of the alien landscape. Most importantly, they could move across the Martian surface. Scientists were no longer stuck in one spot. These rovers traveled across Mars on missions that often lasted for years.

CHAPTER 53
SOJOURNER

- Mission: Mars lander / rover
- Launch: December 4, 1996
- Launch Vehicle: Delta II
- Mars Touchdown: July 4, 1997
- Mission End: September 27, 1997
- Weight (mass) rover only: 23 pounds (10.4 kg)

IN THE MID-20TH CENTURY, THE UNITED STATES AND THE SOVIET Union were locked in the Space Race with each other. In 1971, the Soviet Union tried to land twin probes on Mars. They were part of the Mars 2 and Mars 3 missions. For the first time, these landers included something new: small rovers that could explore the Martian surface.

The Soviet machines were called the Prop-M rovers. About the same size as modern video game consoles, they had moveable metal skis on the bottom. They used the skis to shuffle along the surface. They could also detect and avoid obstacles. Unfortunately, both of the Soviet landers failed. The rovers never had a chance to perform their mission. Mars 2 crashed on the surface, and Mars 3 lost contact with Earth within minutes after landing.

In 1997, the United States returned to Mars. More than two decades had passed since NASA had sent the Viking landers to the Red Planet. The new lander's name was Pathfinder. Piggybacking on Pathfinder was a six-wheeled, microwave-oven-sized rover called Sojourner. (The rover was named after the African American Civil War abolitionist Sojourner Truth. The word sojourner means "traveler.")

Pathfinder was part of NASA's Discovery Program. Its motto was "better, faster, cheaper." Each mission was designed to be built quickly and for relatively little money (for a spacecraft). Pathfinder had a simple mission: test the Sojourner rover in the cold, hostile environment of Mars.

The Sojourner rover measured about two feet (.6 m) long, 1.5 feet (.45 m) wide, and one foot (.3 m) high. It weighed approximately 23 pounds (10.4 kg). The rover had cleats on each of its six wheels to give it traction in the dusty Martian soil. Its top speed was about two feet per minute (.6 m/min).

A solar panel on the rover's roof provided power, while five lasers helped controllers on Earth identify and avoid obstacles. Two small, front-mounted black-and-white cameras took stereo pictures. A color camera was in back. An alpha proton X-ray spectrometer identified the kinds of rocks and soil Sojourner would discover on its mission.

Sojourner got around the Martian landscape with the help of six aluminum wheels. Each had stainless steel treads and cleats for maximum traction, and were five inches (13 cm) in diameter. The robotic rover's rocker-bogie suspension gave the wheels great stability, which was important for the uneven terrain it would encounter. Sojourner could even climb over rocks eight inches (20 cm) high if necessary. Normally, however, it would use a laser system to detect and avoid large obstacles.

NASA engineers standing by duplicates of three generations of Mars rovers. Sojourner was much smaller than the rovers that came after it, but it accomplished much, working 83 days on the Red Planet.

CHAPTER 54
PATHFINDER AND SOJOURNER REACH MARS

On December 4, 1996, the Pathfinder lander, with Sojourner nestled safely within, blasted off from Florida's Cape Canaveral Air Force Station atop a powerful Delta II rocket. The spacecraft began a long, arching path to catch up with Mars on its orbit around the Sun. On July 4, 1997, after a trip of more than 300 million miles (483 million km), it reached its destination. Operating completely on its own, it entered the Red Planet's outer atmosphere at about 16,000 miles per hour (25,750 kph).

Aerobraking through the thin atmosphere helped Pathfinder slow down to about 800 miles per hour (1,287 kph). Protected by a heat shield, the spacecraft blazed a fiery trail across the pink Martian sky.

Several miles above the surface, a huge, 36-foot (11-m) -diameter parachute opened, slowing the spacecraft down to about 160 miles per hour (257 kph). A few seconds later, the heat shield was discarded.

Directly below Pathfinder lay its target: Ares Vallis. It was a wide floodplain at the mouth of a major canyon system. At one time, floodwaters equal to all five of Earth's Great Lakes washed through this now-dry valley. Geologists expected to find all kinds of rocks strewn over the landscape.

Eighty seconds from touchdown, about three miles (4.8 km) from the surface, the main lander was quickly winched down from the rocket pack. It dangled on a 65-foot (20-m) -long tether made of tough Kevlar fabric. The lander used radar to measure its distance from the surface. Eight seconds before landing, sets of large airbags were rapidly inflated that protected the precious cargo inside. Braking rockets fired briefly, and the tether was released. Pathfinder free-fell to the surface, 70 feet (21 m) below. It impacted at a speed of about 40 miles per hour (64 kph).

When the airbag-encased spacecraft hit, it bounced nearly 45 feet (14 m) high, then fell to the ground again. It bounced at least 15 times, finally settling to a stop on the frigid plain of Ares Vallis.

For the first time in 21 years, humans had made their mark on Mars once again. Scientists at NASA's Jet Propulsion Laboratory (JPL), who were overseeing the mission, were overjoyed.

Pathfinder had landed at about 3:00 a.m. on Mars. The airbags were slowly deflated, and the lander's three solar panels opened like a big, metallic flower. When dawn struck, rays from the Sun began powering the probe's instruments.

CHAPTER 55
THE FIRST ROVER TO EXPLORE MARS

AFTER LANDING ON MARS, PATHFINDER SENT ITS FIRST PHOTOS OF the alien landscape that morning. The detailed images were stunning. The scene looked like it was from the American Southwest, but without cacti or sagebrush. The rust-colored Martian soil was littered with rocks and boulders of every shape and size. Geologists on the NASA team were thrilled. In the distance, under a salmon-colored sky, were two tall hills. They were later named Twin Peaks.

NASA announced that it had renamed the lander the Carl Sagan Memorial Station, after the popular astronomer Carl Sagan. He had died earlier in 1997. His work had greatly helped in the push to send spacecraft to Mars. On that first day, the lander sent a weather report. The temperature was a numbing -64 degrees Fahrenheit (-53° C). The spacecraft was expected to withstand such conditions for about 30 Martian days (called sols).

On sol two, it was the Sojourner rover's turn to shine. To move around on the surface of Mars, Sojourner received its instructions from controllers back on Earth. But operating Sojourner was nothing like playing with a radio-controlled car or helicopter. During its mission, Earth and Mars would be 119 million miles (192 million km) apart. Radio signals traveling at the speed of light would take about 11 minutes to get to Mars. The rover's controllers had to give instructions that were slow and deliberate. If something went wrong, they couldn't react in time to save the little rover.

Sojourner rolled down a steep ramp toward the surface of Mars. The process took most of the day. Controllers at JPL were cautious because so many things could go wrong. The small rover could tip, or get a wheel stuck on the edge of the ramp, or even get tangled in the deflated fabric of the lander's airbags.

By the end of sol two, Sojourner had all six wheels planted firmly on Martian soil. Its first task was to sniff the dirt and nearby rocks with its alpha proton X-ray spectrometer. For 10 hours, it sent data to the lander, which relayed the information back to Earth.

In the following days, Sojourner moved to investigate rocks around the landing site, including a boulder called Yogi. Sometimes Sojourner ran across large rocks, but its six metal-cleated wheels moved independently of each other, which usually kept it from getting stuck.

Sojourner analyzed the Martian rocks and soil and took pictures. It never roamed more than about 30 feet (9 m) from the Carl Sagan Memorial Station (Pathfinder), but there was plenty to study. A huge flood had clearly washed across this part of the planet long ago. The rocks were volcanically formed, but many had to have come from other parts of the planet during the flood. But where had all the water gone? That question would have to be answered by future missions.

The Carl Sagan Memorial Station (Pathfinder) was designed to last 30 days. Sojourner was expected to last just seven days. However, both machines kept working for many extra weeks. Finally, on September 27, 1997, the lander sent its last message from Mars. It had fallen victim to the planet's harsh conditions.

By mission's end, the lander had sent back over 16,000 images, plus 550 images from Sojourner. It also sent more than 15 chemical analyses of rocks, plus data about Mars's weather patterns. The mission also proved that the new technology it used could work on Mars. Future spacecraft builders could learn from its success.

CHAPTER 56
SPIRIT AND OPPORTUNITY

- Mission: Mars rovers
- Launch: Spirit: June 10, 2003
- Opportunity: July 7, 2003
- Launch Vehicles: Delta II
- Mars Touchdown: Spirit: Jan. 4, 2004
- Opportunity: Jan. 25, 2004
- Mission End: Spirit: March 22, 2010
- Opportunity: February 13, 2019
- Weight (mass) rover only: 384 pounds (174 kg)

AFTER THE SUCCESS OF THE 1997 PATHFINDER MISSION, NASA prepared to return to the Martian surface. This time, twin rovers would explore the Red Planet. The mission of the Mars Exploration Rovers (MER) program was to look for evidence of water in the planet's distant past. The rovers would do this by analyzing the rocks and soil they encountered. Some rocks and minerals are formed only in the presence of water.

By finding clues to past water, scientists could tell what kind of wet environments existed on ancient Mars. Water is a necessary ingredient for life as we know it. NASA hoped to find out if life could have sprung from Mars long ago.

The two MER program rovers were named Spirit and Opportunity. The names were chosen from a student essay contest that had almost 10,000 entries. Spirit's lander was named Columbia Memorial Station. Opportunity's lander was named Challenger Memorial Station. They were named in honor of destroyed space shuttles Columbia (2003) and Challenger (1986) and the heroic astronauts who died in the tragedies.

The golf-cart-sized rovers were designed and built at NASA's Jet Propulsion Laboratory in Pasadena, California. During their construction, scientists studied high-resolution images of Mars taken by the Mars Global Surveyor and Mars Odyssey orbiters. They chose two landing sites on roughly opposite sides of the planet. Each site seemed promising. The Spirit rover would go to Gusev Crater, a 103-mile (166-km) -wide basin that may once have held a lake. Opportunity would land in a broad plain called Meridiani Planum. Orbiter images hinted it may also have had a watery past.

Spirit and Opportunity borrowed many design ideas from the Pathfinder mission. Each rover had six wheels and used rocker-bogie suspensions. This would allow them to roll over obstacles and climb steep hills. Other similarities to Pathfinder included airbags to cushion landing on the surface and using solar panels for power. The solar panels were mounted on "wings" that unfolded on the top of each rover.

Spirit and Opportunity were about 17 times heavier than the Sojourner rover. Sojourner's host, the Pathfinder lander, housed heavy communications equipment for contacting Earth, plus cameras and computers. Spirit and Opportunity carried all that equipment themselves. Their landers were built to protect them in flight. Once on the ground, the landers would unfold. The rovers would roll down a ramp onto the Martian surface, and the landers' mission would be over.

Spirit lifted off from Florida's Cape Canaveral Air Force Station on June 10, 2003. Opportunity followed on July 7, 2003. Each rover was tightly packed inside its lander. They survived dangerous solar flares during their six-month journeys through deep space.

Spirit was the first to arrive at Mars. On January 4, 2004, the spacecraft hit the Martian atmosphere. As tremendous friction from the air slowed the lander down, a heat shield protected the rover inside. Acting on its own, it released a parachute. A few minutes later, the heat shield was released, and the rover was winched down on a 65-foot (20-m) tether. The lander used radar-guided retrorockets to slow the descent even more. The airbags around the rover inflated, and the tether was cut 40 feet (12 m) from the surface.

Protected by the airbags, Spirit hit the ground. It rolled and bounced 28 times before finally coming to a rest inside Gusev Crater. It was a near-perfect landing. After traveling 303 million miles (488 million km), Spirit was ready to begin its mission.

Right after touchdown, the lander retracted its airbags and opened its petals. Spirit unfolded its solar arrays and raised its camera mast, ready to roll onto Mars.

CHAPTER 57
CRATERS AND DUST DEVILS

After Spirit touched down inside Gusev Crater, the lander's side panels unfolded. Winches and cables retracted the airbags. Spirit used one of the side panels as a ramp to slowly roll down to the planet's surface. Its cameras took pictures of the surrounding area. High-resolution panoramic images amazed scientists on Earth. The landscape was mostly flat and strewn with small rocks.

For the first couple of weeks, Spirit slowly roamed around the landing site. It took photos and analyzed the area's geology. Scientists were disappointed with one of the earliest discoveries. Gusev Crater was filled with volcanic rock, which was mostly basalt. Orbiter images made scientists believe the area was once flooded with water, which is why Spirit was sent there in the first place. If that were true, then later volcanic lava flows had covered up any hint of a life-supporting environment long ago.

A computer glitch struck Spirit during its third week on Mars. As computer experts on Earth worked on the problem, the Opportunity rover arrived. Nicknamed "Oppy," it landed safely at Meridiani Planum on January 25, 2004. After bouncing 26 times on its airbags, it came to rest in a shallow depression called Eagle Crater. Oppy landed just 16 miles (26 km) from the center of its target zone. One scientist called it an "interplanetary hole in one."

Eagle Crater was formed by a meteorite that slammed into the ground. The impact exposed layers of rock on the sides of the crater. The layers appeared to be sedimentary, formed by water sometime in the planet's distant past. Opportunity investigated the rocks with a microscope. It also used spectrometers that identified chemical elements. (Elements react in unique ways to light. That includes visible light, like we see with our eyes, and invisible light, like microwaves and X-rays.)

"Blueberries"

Opportunity also found tiny blue-gray rocks shaped like balls. Thousands of them were strewn across the crater floor. Nicknamed "blueberries," they were made of the mineral hematite. It usually forms in the presence of water. Oppy had discovered more proof of Mars's wet history.

Meanwhile, on the other side of the planet, Spirit's software problems were fixed. It continued exploring the Martian landscape. It used a rock-grinding tool and a microscope to examine the volcanic basalt of the area.

Spirit began a long trip to an area called Columbia Hills (named in honor of the destroyed space shuttle). It climbed hills and found exposed rocks that were probably altered by water, unlike the volcanic basalt in the rover's landing zone.

During its journey, Spirit began losing power. NASA controllers realized that after months of exploring the planet's surface, the rover's solar panels were becoming covered in Martian dust. The fine-grained dust blocked the Sun's rays. Luckily, a dust devil (a small, tornado-like whirlwind) eventually came along and blew much of the debris off the solar panels. The Opportunity rover also had its solar panels cleared by dust devils and windstorms. These occasional cleanings boosted Spirit and Opportunity's power. It was one reason why the rovers were able to explore Mars far longer than expected.

A Martian dust devil photographed by Opportunity.

CHAPTER 58
THE LONG JOURNEY

SPIRIT AND OPPORTUNITY EACH LASTED MUCH LONGER THAN scientists had planned. Their primary missions were scheduled to last just 90 days. In fact, each rover spent many years roaming the Martian landscape. Spirit explored many features inside the Gusev Crater basin. It drove partway around the rim of Bonneville Crater, and climbed to the exposed bedrock atop Columbia Hills.

Opportunity began its own long trek after climbing out of Eagle Crater in mid-2004. Its first major stop: Victoria Crater. One-half mile (.8 km) in diameter, the crater had eroded "fingers" around its rim that would be interesting to study. Along the way, Opportunity drove past its own heat shield, which was jettisoned during landing. Nearby was an intact meteorite embedded in the soil. About the size of a basketball, it was named Heat Shield Rock.

Driving across Mars was a slow process for both rovers. They usually ambled over the surface at just two feet (.6 m) per minute. On a good day, the rovers might cover a few hundred feet of distance, analyzing rocks and taking stunning panoramic photos of the Martian landscape as they went. Each step in the journey was carefully planned by controllers on Earth. Sometimes, when long stretches of roving were planned, Spirit and Opportunity could drive by themselves. When the rovers met unexpected obstacles, they drove around them automatically or stopped and radioed home for help.

Heat Shield Rock meteorite

During winter, the rovers were put into sleep mode at night to conserve energy. Dust gathering on the solar panels reduced their available power. At night, temperatures on Mars often plunged to -140 degrees Fahrenheit (-96° C) or colder. Internal heaters protected the rovers' electronics, but with less power to draw from, the cold eventually took its toll.

In 2009, Spirit's wheels got stuck in sandy soil. Controllers tried for months to free the rover, but it was no use. Then, on March 22, 2010, Spirit sent its last radio signals to Earth. It had spent over six years exploring Mars before succumbing to the harsh conditions. During its service, Spirit took more than 128,000 photographs and uncovered more evidence that water had once covered much of the Red Planet.

In 2008, Opportunity left Victoria Crater and began a long journey to Endeavour Crater. Seven miles (11 km) distant, the trip took almost three years. Along the way, the rover took many side trips to explore and analyze interesting rocks and craters. It also found more meteorites.

In August 2011, Opportunity reached the rim of Endeavour Crater. This was a much bigger crater than Victoria Crater, measuring about 14 miles (23 km) in diameter. Opportunity spent years at the rim because of the interesting rocks and bluffs in the area. Many of the rocks that the rover analyzed could only have formed long ago in the presence of water. This reaffirmed that parts of Mars must have been wet for long periods of time. With so much evidence of water, the possibility of microbial life on the planet long ago cannot be ruled out.

Starting in June 2018, a swirling dust storm enveloped most of the planet. Oppy's solar panels were affected by the dust and became less and less efficient. Finally, on June 10, the rover's power supply sank so low that it could no longer communicate with Earth. It went into "safe mode," a kind of hibernation that shut down non-essential electronics. Months later, after the dust storm finally cleared, it was hoped that Martian winds might clear the solar panels, but scientists never again heard from Oppy, even after hundreds of attempts to contact it. NASA declared Opportunity's mission over on February 13, 2019.

After nearly 15 active Earth years on Mars, Opportunity traveled a record-breaking 28 miles (45 km), traversing rocks, gravel-strewn hills, craters, and what appeared to be ancient dry riverbeds, until it reached its final destination: Perseverance Valley, an apt name for the hardy rover. During its mission, Oppy explored the planet's geology and weather, and took more than 228,000 pictures.

Spirit and Opportunity helped us unravel some of the mysteries of Mars. Their search for the Red Planet's watery past and the possibility of microbial life helped pave the way for NASA's next rover: Curiosity.

A dusty Opportunity rover at the rim of the huge Endeavour Crater.

CHAPTER 59
CURIOSITY

- Mission: Mars rover
- Launch: November 26, 2011
- Launch Vehicle: Atlas V
- Mars Touchdown: August 6, 2012
- Mission End: Continuing
- Weight (mass) rover only: 1,982 pounds (899 kg)

In 2012, NASA landed its most ambitious rover yet on Mars: Curiosity. The space agency's engineers began planning the rover's mission in the early 2000s. Its most important goal was to discover if Mars ever had the right conditions to support life. It would do this by analyzing the Martian soil, rocks, and air. The rover would especially look for organic material. Organics are the chemical building blocks of all life as we know it.

As part of the Mars Science Laboratory mission, the Curiosity rover took more than seven years to plan, build, and test. About the size of a compact car, the rover has six wide wheels with a rocker-bogie suspension, similar to its predecessor rovers: Spirit, Opportunity, and Sojourner. It allows Curiosity to roll over tall obstacles and climb steep hills. Unlike those previous rovers, however, Curiosity was built much bigger so that it can hold more scientific instruments. It measures 10 feet (3 m) long, 9 feet (2.7 m) wide, and 7 feet (2.1 m) high. It weighs nearly one ton, tipping the scales at 1,982 pounds (899 kg). It is about twice as long and five times as heavy as either the Spirit or Opportunity rovers.

Previous Martian rovers got their power from solar panels. However, dust often gathered on the flat panels, which reduced the amount of sunlight that could generate electricity. Instead of solar panels, Curiosity has a radioisotope power system. It uses the heat generated by radioactive plutonium-238 and converts it to electricity. It is very dependable and provides more than enough power for roaming across the Martian landscape and for all the rover's scientific instruments.

Two identical computers control Curiosity. If one computer crashes or has other problems, the backup computer takes over. Curiosity can directly communicate by radio with controllers on Earth. It can also send data to satellites orbiting Mars, which then relay the messages to Earth.

Curiosity bristles with scientific equipment. It has spectrometers and lasers that analyze the chemical elements in rock, soil, and air samples. It has a radiation detector. It also has 17 cameras. Some take high-resolution photos of the Martian landscape, while others help avoid hazards while roaming. Several are mounted on a tall mast that extends several feet upward.

The rover is also equipped with a 6.9-foot (2.1-m) -long robotic arm. There are several instruments mounted on the end. They include a close-up camera, X-ray spectrometer, drill, rock-cleaning brush, and a scoop. After drilling .6-inch (1.5-cm) -wide holes in rocks or soil, the arm gathers up the powdered samples and moves them to chemistry labs inside the rover.

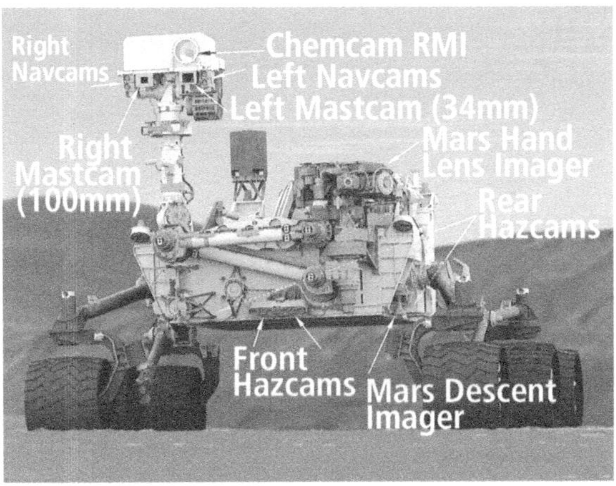

Curiosity carries navigation, scientific analysis, and photo cameras.

CHAPTER 60
SEVEN MINUTES OF TERROR

THE CURIOSITY ROVER WAS CARRIED TO MARS ABOARD THE MARS Science Laboratory (MSL) spacecraft. The MSL was equipped with computers, rockets, and fuel. A 14.8-foot (4.5-m) heat shield would protect the rover during the trip through the Martian atmosphere, and a large parachute would help slow its speed. The MSL was assembled at NASA's Jet Propulsion Laboratory in Pasadena, California.

The MSL spacecraft, with Curiosity aboard, launched from Florida's Cape Canaveral Air Force Station on November 26, 2011. It was carried into space by a powerful Atlas V rocket. During its eight-month trip to Mars, the spacecraft traveled about 352 million miles (566 million km).

Curiosity landed on Mars on August 6, 2012. Getting to the planet's surface was a tense ordeal. Because of the millions of miles of distance between the two planets, Curiosity had to land automatically. Everything had to work just right and be perfectly timed. If any one piece of equipment failed, the spacecraft would crash.

It would take seven minutes for the spacecraft to travel from the top of the atmosphere to the surface. It took 14 long minutes for the rover's radio signals to reach Earth. For that reason, by the time NASA scientists got word that Curiosity had entered the top of the atmosphere, the rover would have already landed safely on Mars, or slammed into the rocky ground and been destroyed. NASA called this the "seven minutes of terror."

About 78 miles (126 km) above Mars's surface, the spacecraft slammed into the atmosphere. Traveling about 13,000 miles per hour (20,921 kph), the friction of speeding through the air caused the heat shield to reach 3,800 degrees Fahrenheit (2,093° C). Inside, Curiosity stayed cool at about 50 degrees Fahrenheit (10° C).

The spacecraft fired small rockets to guide it to its landing zone. About seven miles (11 km) above the ground, the spacecraft's speed dropped to 900 miles per hour (1,448 kph). Soon, a large parachute billowed out, and the heat shield popped off.

Radar helped the spacecraft judge the distance to the surface. About one mile (1.6 km) up, the top shell and parachute were ejected, and the spacecraft went into free fall, traveling 170 miles per hour (274 kph). Eight retrorockets fired, slowing the descent. In the final few seconds, the spacecraft came to a near-hover.

Unlike previous rovers, Curiosity was too heavy to safely land using airbags to cushion the impact. Instead, it used a radical new method called a "sky crane." The rover was lowered from the upper-stage spacecraft to the ground by three 21-foot (6.4-m) - long nylon ropes. This kept the rover far away from the spacecraft's retrorockets. Otherwise, they would have stirred up dust and rocks on the Martian surface, which could have harmed the rover.

After Curiosity safely touched down, the ropes were cut and the upper stage flew off and crashed about 500 feet (152 m) away so it wouldn't harm the rover. Curiosity signaled its controllers on Earth: it was ready to begin its mission.

Once Curiosity was safely lowered to the ground, the sky crane ropes were cut free and the upper-stage spacecraft flew off and crashed a safe distance from the rover.

CHAPTER 61
EXPLORING GALE CRATER

CURIOSITY MADE AN ALMOST pinpoint landing inside Gale Crater, in the Elysium Planitia region of Mars. The site was named Bradbury Landing, in honor of author Ray Bradbury. Gale Crater had been chosen because there is evidence that ancient waterways once flowed there. A meteorite slammed into Mars billions of years ago, forming the 100-mile (161-km) -diameter crater. It later filled with water. Over millions of years, layers of sediment and erosion formed a large bulge that rises 18,000 feet (5,486 m) in the middle of the crater, called Mount Sharp. Now that water has vanished from the surface of Mars, scientists hope that Curiosity will uncover clues about the planet's watery past.

Curiosity began sending home spectacular high-resolution photographs of its surroundings. Within a few days, NASA controllers sent the rover instructions to begin exploring the area. It zapped nearby rocks with a laser and analyzed the vapor with a spectrometer to figure out their type. It also collected many soil and bedrock samples with its drill and scoop.

In July 2013, Curiosity began a long journey to the base of

Mount Sharp. Along the way, it took jaw-dropping images of the Red Planet, and studied many interesting rocks. It arrived at the base of Mount Sharp in September 2014.

In June 2018, NASA announced that Curiosity had discovered organic molecules on the lower slopes of Mount Sharp. They were preserved inside three-billion-year-old sedimentary mudstone (made from muds and clays). The chemistry lab inside Curiosity heated powdered rock samples to 932 degrees Fahrenheit (500° C). This released chemicals in the form of gas so they could be analyzed. Organic molecules contain carbon atoms and are the basic building blocks of living creatures, including microbes.

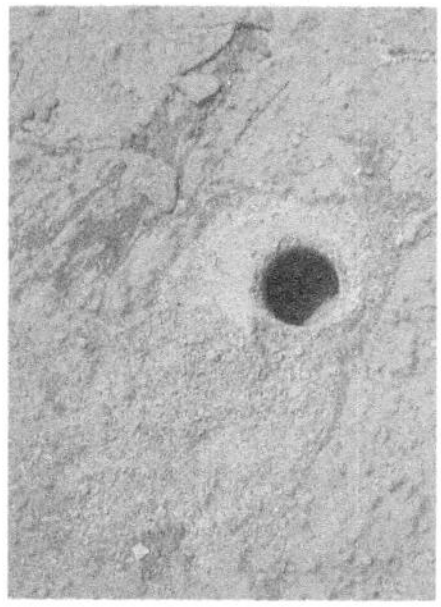

A drill hole

By August 2022, Curiosity had climbed over 2,050 feet (625 m) up the slopes of Mount Sharp and analyzed 41 rock samples from different sedimentary layers. Each geologic layer tested so far had the capacity to harbor life, a time period spanning millions of years. This long period of habitability surprised scientists.

Curiosity also discovered methane gas in the Martian air. The rover "breathes" the Martian atmosphere through two intake valves on its side and analyzes it. Scientists discovered that the levels of methane increase seasonally. Methane is often found as a byproduct of living organisms. Scientists wonder if the increasing methane is coming from active colonies of underground microbes during the warmer summer months.

Curiosity's robotic arm is seen as the rover explores the Yellowknife Bay basin inside Gale Crater. Mount Sharp rises in the background.

Has Curiosity discovered proof of life on Mars? Not yet. Even though it's possible microbes are releasing methane into the air, both the ancient carbon molecules in the rocks and methane in the air could come from other natural sources. More exploration is needed. Curiosity and future rovers will look for more evidence that could prove creatures live on the Red Planet.

What Curiosity *has* proved, however, is that Mars was once habitable. It *could* have harbored life for millions, and perhaps tens of millions, of years. The planet has been explored enough today to all but rule out larger life forms. There are no dinosaur footprints to be seen, and no complex animal fossils. Perhaps future rovers will prove this wrong, but if life *did* exist on Mars, it probably got no further than the microbial stage.

But why? We know from data and images collected by orbiting satellites that early Mars—three to four billions of years ago—was a more Earth-like planet. It had a warmer atmosphere and flowing rivers of water, perhaps even an ocean. But Mars is much smaller than Earth. It cooled quickly and lost most of its magnetic field, which led to space radiation drastically thinning its atmosphere. Without the ability to stay warm and harbor liquid water, the chances of developing complex life forms as we know them faded away.

Curiosity was built to last two years, but its mission has lasted far longer. As of August 6, 2022, it has been exploring Mars over five times its expected life span. It has survived frigid Martian winters and traveled more than 18 miles (29 km) across the alien landscape. Along the way, it has uncovered further evidence that Mars was once a wet planet, with conditions that could have harbored microbial life in the distant past. With any luck, Curiosity will continue its exploration for many years to come.

For updates and more information about Curiosity, visit NASA's website at: https://mars.nasa.gov/msl/home/

A selfie from Curiosity. The rover is dusty and showing signs of wear on its wheels, but it continues exploring the Red Planet far beyond its expected life span.

CHAPTER 62
PERSEVERANCE

- Mission: Mars rover
- Launch: July 30, 2020
- Launch Vehicle: Atlas V
- Mars Touchdown: February 18, 2021
- Mission End: Continuing
- Weight (mass) rover only: 2,260 pounds (1,025 kg)

In the summer of 2020, an Atlas V rocket lifted off from Florida's Cape Canaveral Space Force Station. It rode of pillar of flames into the morning sky, quickly leaving the bonds of Earth's gravity and entering the inky blackness of space. Safely tucked away inside was Perseverance, the flagship of NASA's Mars 2020 mission. Its primary goal: to search for signs that microbial life once existed on Mars.

On February 18, 2021, after a seven-month journey through deep space, Perseverance landed in Jezero Crater, near the Martian equator. It employed a sky crane landing system like its predecessor, Curiosity. After an uneventful touchdown inside the 28-mile (45-km) wide crater, Perseverance began its mission, which is scheduled to last at least one Martian year (687 Earth days). The landing site was named Octavia E. Butler Landing, after the acclaimed American science fiction author.

Jezero Crater was chosen by NASA because it was once filled with water for millions of years. A large fan-delta formed near an inflow channel of the lake. Scientists surmise that if life developed during this period, there may be signs of it locked away in sedimentary rock deposits. On Earth, microscopic fossils can be found preserved in carbonates and hydrated silica. If the Perseverance rover, nicknamed Percy, is to find evidence of ancient life on Mars, Jezero Crater is an excellent place to start.

Although the Perseverance and Curiosity rovers appear identical at first glance, Percy is slightly bigger and more technologically advanced. Its wheels are tougher, its computers are more powerful, and it has a stronger sampling arm. It carries a drill to collect core samples of Martian rocks, and a radar unit to probe down to 30 feet (9.1 m) below the surface in search of ice, water, or other buried objects of interest. It uses magnifying cameras, spectrometers, and a laser to detect organic compounds and other potential signs of ancient microbial life. The rover has 19 cameras in total, and is also equipped with microphones to capture the sounds of Mars.

Perseverance is capable of driving .1 mile per hour (160 meters per hour) across flat Martian terrain. It must slow down when faced with obstacles. However, instead of controllers on Earth telling it every move to make, which takes several minutes for each command, Percy can detect obstructions in its path and maneuver around them on its own. This allows the rover to travel autonomously toward science goals such as rocks from which to extract core samples. The rover generates maps of the terrain while driving and selects the safest and most efficient routes to proceed.

Also helping Percy drive across Mars is a companion named Ingenuity. It is a small, robotic helicopter with four carbon-fiber blades and a solar panel that generates power for flights lasting about 90 seconds. It weighs less than 4 pounds (1.8 kg). Stowed to the rover's underbelly, it was dropped off in order to perform a series of tests in the thin Martian air. On April 19, 2021, it climbed about 10 feet (3 m) into the air, hovered, and then settled back to the ground, completing the first powered flight of an aircraft on Mars, or any planet other than Earth.

As of September 24, 2022, Ingenuity has flown 33 times, for a total distance of 24,253 feet (7,392 m). Although not designed to support Percy's mission, Ingenuity has been so successful that NASA started using it to scout locations for the rover. Future Mars helicopters may someday not only help find places for rovers to study, but may also explore cliffs and caves that rovers cannot reach.

NASA's Mars 2020 rover will explore the Red Planet and test the Mars helicopter Ingenuity.

As of October 2022, Perseverance has traveled over eight miles

(13 km) and gathered 14 samples of Martian rock, regolith (Martian soil), and air, representing a diverse collection from Jezero Crater's ancient river delta. The rover has detected organic molecules that, under the right circumstances, are the chemical building blocks of life. But in order to tell if life actually existed on Mars, more in-depth experiments will need to be conducted on the rock samples, tests that the rover is incapable of performing.

Scientists have long believed that to examine Martian rocks in the most complete way possible, the samples must be transported to science labs on Earth, where advanced microscopes, spectrometers, and other instruments can analyze them, and perhaps even spot microfossils! That is why a major part of Perseverance's job is to cache rock core samples in sealed containers, about the size of a pinkie finger, for a future mission to transport them back to Earth.

As of November 2022, the plan for the Mars Sample Return mission involves several spacecraft working together. A separate lander called the Sample Retrieval Lander will arrive on the surface somewhere close to Perseverance in Jezero Crater. Percy will approach the lander and begin unloading its supply of core sample containers. A robotic arm—developed by the European Space Agency—attached to the lander will load each container into a small rocket called the Mars Ascent Vehicle.

If something goes wrong with the transfer procedure, there is a backup in place. Before the lander arrives—perhaps as early as November 2022—Perseverance will cache a duplicate set of sample tubes on the surface of Mars. The drop-off location—called the sample depot—will be near the base of the river delta at a place called Three Forks. Two robotic helicopters, their design based on Ingenuity, will ferry the sample containers from Three Forks to the Sample Retrieval Lander.

Once the sample tubes are securely loaded, the rocket will launch into Mars orbit. There, it will rendezvous with yet another spacecraft, called the Earth Return Orbiter. After the container of sample tubes is transferred, the spacecraft will then make the long voyage back to Earth. A disk-shaped craft with the samples inside, called the Earth Entry System, will separate from the Earth Return Orbiter and make a fiery descent through Earth's atmosphere, safely touching down on land with its precious cargo. If all goes according to plan, NASA hopes to have Martian samples in Earth labs by 2033.

For updates and more information about Perseverance, visit NASA's website at: https://mars.nasa.gov/mars2020/

The Perseverance rover will transport Martian rock, soil, and air samples to the Sample Retrieval Lander, handing off a collection of sample tubes carried on board. A pair of Ingenuity-like helicopters will transport the tubes if something goes wrong with Ingenuity. An ESA-provided robotic arm will transfer the tubes into a sample container on the lander's rocket.

CHAPTER 63
ZHURONG

- Mission: Mars rover
- Launch: July 23, 2020
- Launch Vehicle: Long March 5
- Mars Touchdown: May 14, 2021
- Mission End: Continuing
- Weight (mass): 530 pounds (240 kg)

Reaching the surface of Mars with a fully functioning spacecraft is a daunting scientific task, and from the 1970s onward, only one country has been successful: the United States. (In 1971, the Soviet Union's Mars 3 lander touched down on the planet, but communication with Earth lasted just seconds and resulted in a partial television image before all contact was lost.)

America's fleet of Martian landers and rovers got some

company on May 14, 2021, when China landed Zhurong, a six-wheeled rover designed to study Martian geology and weather. It touched down safely in Utopia Planitia, the same large, flat basin visited by NASA's Viking 2 lander from 1976-80.

Zhurong is named after a fire god in Chinese mythology (in China and other Asian countries, Mars is referred to as the Planet of Fire). The rover is part of the China National Space Administration's Tianwen-1 (translation: "Heavenly Questions") mission, China's first interplanetary undertaking.

Zhurong was carried to Mars attached to an orbiter. After three months circling the planet, the rover hitched a ride to the surface aboard a lander. A parachute and retrorocket helped the lander achieve a soft touchdown. After several days of preparations and systems tests, the rover rolled off the lander onto the Martian surface on May 22, 2021.

Zhurong's mission is to map its landing zone and study the geology found there, as well as transmit regular weather reports and search for signs of water ice. Roughly the size of NASA's Spirit and Opportunity rovers, the solar-powered Zhurong is equipped with cameras, ground-penetrating radar, a magnetic field detector, weather instruments, and a laser than can analyze rock samples from a short distance. Data from the rover is beamed to the Tianwen-1 orbiter overhead, and then relayed back to Earth.

By analyzing the chemical composition of nearby rock samples, Zhurong has revealed how ancient water and wind erosion have altered the local geology. The data sheds light on how the history of Mars's water and climate evolved. In addition, the rover's ground-penetrating radar may have found evidence of two catastrophic floods that swept over Utopia Planitia millions of years ago.

Zhurong has far surpassed its original 90-days of operation. As of November 2022, it continues moving southward, exploring, mapping, and collecting data as it roams the Martian landscape.

CHAPTER 64
FUTURE ROVERS

THE EUROPEAN SPACE AGENCY'S EXOMARS ROVER LOOKS SIMILAR TO NASA's Spirit and Opportunity rovers. Its primary mission will be to find signs of life on Mars. The rover's science equipment includes ground-penetrating radar, advanced spectrometers for analyzing rocks, and chemistry labs to look for organic molecules. Its drill can burrow almost seven feet (2.1 m) underground to find unspoiled rock core samples. The ExoMars rover mission was originally a partnership with Russia's Roscosmos space agency, but because of the Russian invasion of Ukraine in early 2022, the project has been suspended indefinitely while the ESA searches for another partner. It could be several years before the mission is restarted.

More rovers and orbiters are planned for the near future. If all goes according to plan, Mars will have a fleet of robotic spacecraft exploring its surface. The data gathered by these scientific marvels will pave the way for human exploration of the Red Planet in the coming decades.

HUMANS TO MARS

CHAPTER 65
THE LURE OF THE RED PLANET

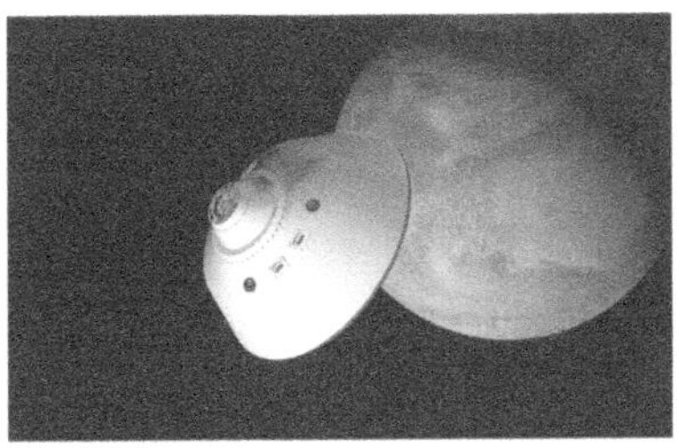

For centuries, people have gazed at Mars, that bright red dot in the night sky, and wondered if there were aliens there. Were they like us? Could we learn from each other? Or maybe the aliens were hostile monsters bent on Earth's destruction. Mars has been the subject of hundreds of science fiction tales, including books and movies. From *The War of the Worlds* to *The Martian*, these stories have set our imaginations on fire.

In the last several decades, the United States and a handful of other countries have sent a fleet of orbiters, landers, and rovers to Mars. We've learned much about the Red Planet. No longer a place of pure imagination, Mars is an actual location that we can explore. As our knowledge and technology grows, the question naturally arises: can people someday set foot on Mars? Can we live and work there in cities? Is it possible to transform that hostile, alien planet so that it is like our own Earth?

These questions are no longer flights of fancy. We have the desire and the technology to colonize Mars. The journey will be filled with danger, and there are many problems to be solved. Despite the hard road ahead, scientists around the world today are working to solve these roadblocks. The future is now.

CHAPTER 66
CAN HUMANS LIVE ON MARS?

THE FIRST QUESTION WE NEED TO ANSWER ABOUT BUILDING COLONIES on Mars is: can humans even live on that remote, forbidding planet? "Absolutely," says Jim Green, a chief scientist at NASA. Green points out that recent discoveries show there is a good chance Mars once harbored life. In fact, microscopic life may live there today.

Probes sent to Mars, such as NASA's Mars Reconnaissance Orbiter and Curiosity rover, have uncovered clues about the planet's ability to support living things. There is much water ice buried under the Martian surface and at the poles. Water is a crucial ingredient for all life as we know it. Mars was once filled with oceans of liquid water. The proof is in the canyons and dried lakes and rivers we can observe today from orbit. Perhaps all that water helped life spring forth, much as it did on Earth.

In fact, there may be microbial life thriving underground on Mars today. Curiosity has detected plumes of methane gas, which is often a byproduct of living organisms. There is no absolute proof yet, but these discoveries cause many scientists to believe that Mars isn't the barren, dead world we once thought it was.

CHAPTER 67
WHY SHOULD WE GO?

THE DISTANCE FROM EARTH TO MARS IS ENORMOUS: 35 MILLION MILES (56 million km) when the planets are closest to each other. A one-way voyage would take at least six to eight months using today's rocket technology. Once on Mars, astronauts will find harsh conditions. It is freezing cold, there is little oxygen to breathe, dust storms regularly blanket the planet, and harmful radiation from the Sun bombards the surface.

Before we build human colonies on Mars, we have to ask ourselves: what are the reasons for going? Can robotic spacecraft alone do the job? If we send astronauts, why should we risk their lives?

One reason to go is simply to learn. Humans on Mars could perform experiments that would take probes much longer. The more we learn about Mars, the more we learn about planets in general, including Earth. Perhaps something we learn about the Red Planet will help us save our own home world someday.

Another reason to go is economic. Perhaps there are minerals on Mars that are scarce on Earth. Or maybe Mars could be used as a base to mine nearby asteroids.

All the difficulties of getting to Mars can be overcome, but it will take great effort and ingenuity. We have faced similar challenges before. In the 1400s, we knew how to sail, but we couldn't get across the oceans. That changed as technology and our knowledge improved. The same will be true of interplanetary voyages.

A scene from the 2015 movie *The Martian*.

CHAPTER 68
TRAINING FOR MARS

In December 2017, President Donald Trump changed the National Space Policy. As a result, NASA's current plan is to send people to Mars by about 2040. The Moon will first be used to test technology and to set up an outpost. The base will either be the Moon itself, or a Moon-orbiting space station called the Lunar Orbital Platform. It will be like the International Space Station (ISS), but smaller. After the Moon is explored and technology for interplanetary travel has been tested, crewed missions to Mars will begin.

In the coming years, NASA will continue exploring Mars with unmanned orbiters, landers, and rovers. This includes a mission to send samples of Mars rocks to Earth. The technology used to make the return trip will help scientists plan for human expeditions.

Outside of the NASA HI-SEAS Mars training dome in Hawaii.

As new spaceflight technology is built, astronauts will train to live on Mars. NASA's Hawaii Space Exploration Analog and Simulation (HI-SEAS) project tests astronaut psychology. Crews of six volunteers spend up to one year inside a domed habitat. The dome was built on the remote foothills of Hawaii's Mauna Loa volcano. The rocky terrain is similar to what is found on Mars. Volunteers are confined inside the dome, where they eat, sleep, exercise, and work. They eat mostly freeze-dried food, but also grow some of their own vegetables. They sometimes go outside, but only when wearing full spacesuits. NASA hopes to learn how future Mars astronauts will get along with each other during missions that can last years.

CHAPTER 69
GETTING TO MARS

GETTING SAFELY TO MARS IS NEVER EASY. ABOUT HALF OF ALL spacecraft sent to the Red Planet have crashed or been sent speeding into deep space by accident. Putting humans inside spacecraft adds another whole level of complexity—and danger.

The biggest problem with sending astronauts from Earth to Mars is the tremendous distance between the two planets. The greater the distance, the longer the trip is going to take.

NASA's Orion spacecraft.

Mars and Earth are constantly moving around the Sun. Their distance apart is always changing. Earth orbits the Sun once every 365 days. Mars takes about 669 Earth days to finish its orbit. That is a little less than two Earth years. Spacecraft can only begin a trip to Mars when the planets are properly lined up, which happens every 26 months.

When they are closest, Earth and Mars are about 35 million miles (56 million km) apart. However, the actual travel distance a spacecraft must travel is much longer. Because the planets are always moving, a spacecraft follows a long arc to complete the trip. The spacecraft must be aimed where Mars will be at the end of the trip, not where the planet is at launch. (This is like a quarterback passing a football to a running receiver.) Because of this, the spacecraft travels about 300 million total miles (483 million km). That's like taking a one-way trip to the Moon 1,256 times!

Using today's rocket technology, the long voyage to Mars takes seven to eight months. That is a very long time for astronauts to live and work in a cramped space. The long travel time also means astronauts will have to bring much food, and recycle their water and air. If something breaks, they will have to fix it themselves with spare parts they bring.

Making the trip in less time would have many advantages. The crew would need to bring fewer supplies. There would be less time for machinery to break down. It would also mean less time that the crew would be exposed to cancer-causing cosmic rays.

The trip to Mars could take less time if a spacecraft could burn more fuel to gain speed. But a mission with a human crew would be many times heavier than a robot probe, especially with all the extra life-support equipment needed. The more weight the spacecraft carries, the more fuel is needed to burn. The extra fuel itself adds even more weight. Bigger rockets with more fuel can be used, but there is a size limit using current technology.

Future spacecraft might use powerful lasers, electromagnetic thrusters, or even nuclear reactors to gain speed. For now, however, NASA will continue to rely on rockets that use chemical reactions to provide thrust. The technology has been used for decades and is dependable. New, faster rocket technology will eventually be used, but it will take time to test and build.

CHAPTER 70
THE ORION SPACECRAFT

Orion is NASA's next generation of manned spacecraft. Named after the constellation in the night sky, Orion will take a crew of four astronauts to the Moon and into deep space. Sometime in the 2030s or 2040s, it will be used to send the first humans to Mars.

For the eight-month voyage to Mars, Orion will include an attached Deep Space Habitat. The bus-sized module is where the four-person crew will live and work safely in deep space. Food, water, and other cargo can also be brought along inside the module.

Besides NASA, private companies are making plans to go into space, and eventually Mars. They include companies such as Boeing, Virgin Galactic, and Blue Origin. SpaceX is a company run by billionaire Elon Musk. In 2017, Musk announced plans to send spacecraft to Mars in the 2020s. His goal is to eventually colonize the Red Planet. That is an ambitious plan, but SpaceX has been very successful so far. Its Falcon-series rockets and Dragon spacecraft have successfully put satellites into Earth orbit and sent supplies to the ISS.

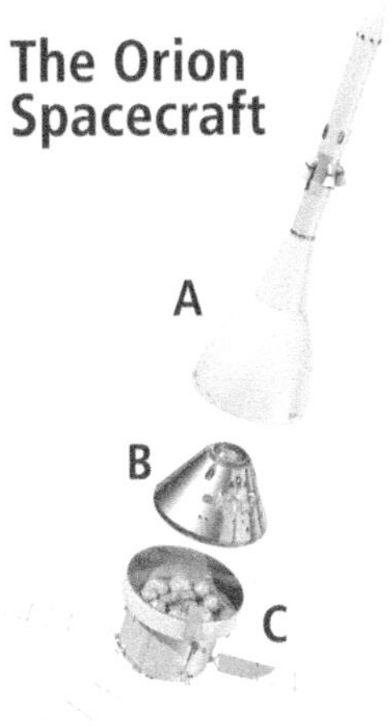

A) LAUNCH ABORT SYSTEM

This solid rocket system is attached above the Orion crew module. In an emergency, the rocket can fire within milliseconds to quickly propel the astronauts out of harm's way. It then moves to a safe position and drops the crew module for a safe landing. If Orion reaches orbit safely, the launch abort system is discarded.

B) CREW MODULE

The Orion crew module is where the four astronauts sit during liftoff and when the capsule returns to Earth. It is similar in appearance to the Apollo spacecraft, but slightly larger. It also has many improved features. They include advanced electronics, navigation, and safety systems. The bottom of the capsule is a heat shield that will protect the crew from the extremely high temperatures of reentry into Earth's atmosphere, which can reach 4,000 degrees Fahrenheit (2,204° C).

C) SERVICE MODULE

The service module is a section that attaches to the rear of the crew module. Built with the help of the European Space Agency, it has engine thrusters for navigating through space. The service module also helps the crew survive by providing water, power, air to breathe, and heat. There are four solar arrays attached that unfold after launch. The service module can also hold cargo that doesn't need to be in the pressurized environment of the crew module. Just before reentry into Earth's atmosphere, the service module is discarded.

CHAPTER 71
SPACE LAUNCH SYSTEM

THE ORION SPACECRAFT WILL BE SENT ON ITS WAY TO MARS BY NASA's Space Launch System (SLS). It is the first large rocket system built since the Saturn V rockets, which boosted the Apollo spacecraft to the Moon. The Space Launch System is about the same size as the Saturn V, but provides much more power. It can boost three times as much cargo weight as the space shuttles, which were retired in 2011.

The Space Launch System has three main parts. The central core section is the biggest. It uses four massive engines. They are powered by 730,000 gallons (2.8 million l) of super-cooled liquid oxygen and liquid hydrogen fuel. The fuel is mixed together and ignited.

Two solid rocket boosters are attached to either side of the core rocket. They give extra power that is needed to lift heavy loads into space. Once ignited, the rockets burn five tons (4.5 metric tons) of solid fuel per second. When the fuel is spent, the boosters fall away from the core section. When the core reaches orbit, it separates from the Orion spacecraft and falls back to Earth.

Once fully assembled with the Orion spacecraft, the version of the Space Launch System used to launch astronauts to Mars will weigh about 6.5 million pounds (2.9 million kg). That is about the same as 10 fully loaded 747 aircraft. It will be able to lift 286,000 pounds (129,727 kg) of cargo into space, about the same as 22 fully grown elephants. The first version of the Space Launch System and Orion spacecraft will start test flights into space starting in the early 2020s.

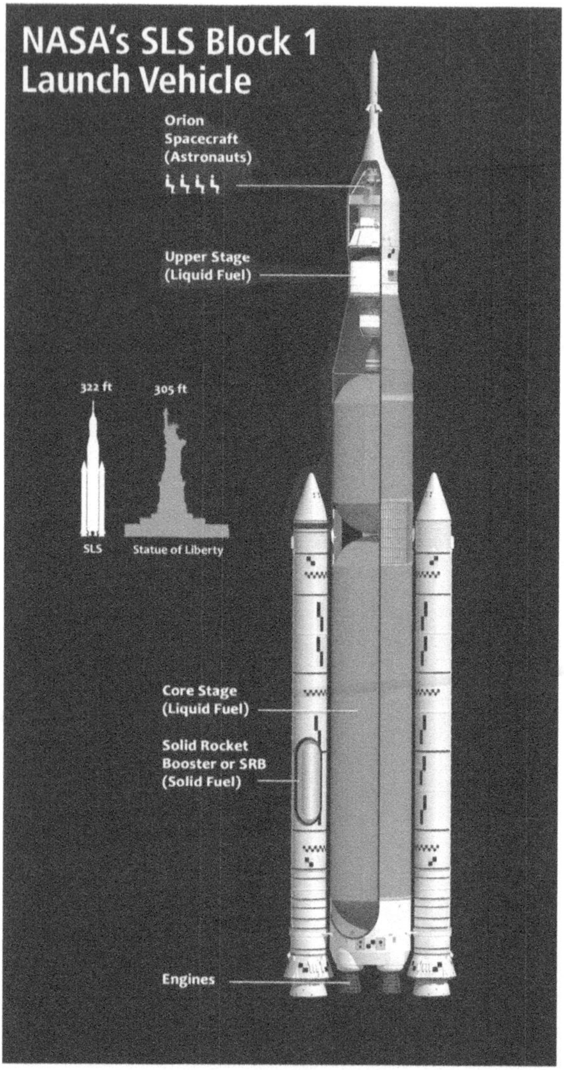
NASA's SLS Block 1
Launch Vehicle
Orion
Spacecraft
(Astronauts)
Upper Stage
(Liquid Fuel)
322 ft
305 ft
SLS
Statue of Liberty
Core Stage
(Liquid Fuel)
Solid Rocket
Booster or SRB
(Solid Fuel)
Engines

CHAPTER 72
THE LONG VOYAGE TO MARS

SURVIVING THE EIGHT-MONTH TRIP TO MARS WILL REQUIRE A LARGE spacecraft. It will need to carry enough food and water to keep the four-person crew alive. It will also have to carry fuel, shield the astronauts from harmful cosmic radiation, and give them enough room to live comfortably during the long journey.

PROPULSION

Spacecraft that use chemical reactions to generate thrust will probably be used on the first interplanetary missions. Some fuel could be made on Mars so the spacecraft wouldn't have to carry a whole round-trip supply. Other options include using nuclear reactors to heat and expel hydrogen, or solar electricity that energizes gasses that create thrust.

LIVING QUARTERS

The trip to Mars is too long for the four astronauts to live in Orion. A Deep Space Habitat module, similar to that shown here, could be assembled in space ahead of the primary mission. The

Orion crew vehicle would attach to the habitat module in orbit, and then the two spacecraft would travel together to Mars.

MENTAL HEALTH

The crew module needs to be big enough so the astronauts are comfortable and their mental health stays positive.

RADIATION

"Green walls" filled with water and edible plants could help shield the crew from harmful cosmic radiation.

REPAIRS

Critical navigation systems and machines that recycle air and water need to be easy to access so the astronauts can repair them if needed.

EXERCISE

Zero gravity takes a toll on the human body. Exercise helps muscles and bones stay in shape and also boosts mental health.

CHAPTER 73
LANDING

When a spacecraft arrives at Mars, it will be traveling many thousands of miles per hour. It needs to slow down somehow. It could reverse its position and fire its engines, but that would use too much fuel. A better option is to use the Martian atmosphere. Even though it is very thin compared to Earth, it is still dense enough to create friction when a fast-moving spacecraft travels through it. The friction will cause the spacecraft to slow down, but will also create scorching heat. The astronauts need to be shielded inside.

After skimming through the upper atmosphere, the spacecraft will slow down and then shoot past Mars. It will go into a wide, elliptical orbit, returning to skim through the atmosphere again and again. Each time, it reduces its velocity even more, until it is moving slowly enough to achieve a circular orbit close to the planet.

Once in a safe orbit, the astronauts will enter a landing vehicle that separates from the main spacecraft. It will protect them from the heat that comes from speeding through the atmosphere. The friction will slow down the landing vehicle, but more help will be needed.

Previous unmanned probes used devices such as airbags to land safely on Mars. Unfortunately, a mission with people and all their supplies will be many times heavier. Old landing methods will not work. A crewed landing vehicle will need parachutes to partially slow down. Then engines will need to fire under the craft so it can land gently. If all goes according to plan, there will be enough fuel left so the astronauts can begin the long journey home.

CHAPTER 74
LIVING ON MARS

MARS IS A HOSTILE PLACE FOR HUMAN BEINGS. THE AIR IS TOO THIN. There is almost no oxygen. Sunlight is weak. Temperatures can sink lower than -100 degrees Fahrenheit (-73° C), and cancer-causing solar radiation constantly bombards the dusty surface. If astronauts hope to survive on the Red Planet, they'll have to bring plenty of supplies with them.

Several companies are testing Deep Space Habitats for NASA that could someday be used on Mars. The companies include Bigelow Aerospace, Boeing, and Lockheed Martin. The habitats range from inflatable fabric shelters to hard-sided structures. For use on Mars, the habitats must be pressurized and sealed in order to provide an Earth-like atmosphere and to keep out dust. The walls must shield against solar radiation. The habitats must also have heating and cooling systems, reliable power systems, docking ports, airlocks, and fire safety equipment.

How much room do astronauts need? That question is more important than it seems. They will be cooped up in one place for many months at a time, with only a few fellow crew members for company. After studying astronauts for many years aboard the International Space Station, NASA discovered that astronauts need more room to feel comfortable and productive.

Inflatable fabric shelters built by Bigelow Aerospace.

All of these requirements for habitats mean several trips will probably be needed to get all the required equipment to Mars. Several loads of heavy gear could be sent ahead of time. Perhaps the habitats and other machinery could be assembled by robots. When a human crew arrives, their base camp will be ready and waiting for them.

CHAPTER 75

A BASE ON MARS

Living on Mars will require overcoming many obstacles. There is very little oxygen, temperatures are extremely cold, there is no water on the surface, and gravity is low. Future astronauts will need smart planning, high-tech equipment, and ingenuity to survive.

NUCLEAR POWER

Reliable energy could be created in a small nuclear reactor that splits atoms. A radiation shield would surround the reactor.

SOLAR POWER

Energy from the Sun could make electricity, even though the Sun's rays are weaker on Mars than on Earth.

LAVA TUBE HOUSE

Underground homes inside volcanic lava tubes can help protect against harmful solar radiation.

OXYGEN

After running electricity through carbon dioxide that is captured from the Martian atmosphere, the molecules split into oxygen and carbon monoxide. The oxygen can be used for breathing, and the carbon monoxide vented back to the outside air.

METHANE GAS

Carbon dioxide and hydrogen can be combined to form water and methane gas. Methane can be used as rocket fuel.

WATER FROM THE AIR

The air on Mars is 95 percent carbon dioxide, but it also contains water vapor. A mineral called zeolite absorbs moisture. The water can then be extracted from the zeolite using microwaves.

GROUNDWATER

Microwaves can be used to melt and extract water ice that is found under the Martian surface.

COMMUNICATION

Messages to and from Earth will be relayed to satellites orbiting Mars.

CHAPTER 76
HOW SPACE AFFECTS THE BODY

SPENDING A LONG TIME IN SPACE AFFECTS THE HUMAN BODY IN extreme ways. Bones and muscles begin to waste away. Astronauts lose about one percent of their bone mass per month. Regularly lifting weights helps, but not totally. Astronauts often exercise as much as two to four hours a day. Many astronauts also suffer a partial loss of vision. Body fluid collects in the brain while in weightlessness, and the fluid presses on their eyeballs.

After working for months in space, astronauts have a tough time coming back to Earth and dealing with gravity again. Astronaut Chris Hadfield spent nearly five months as commander of the International Space Station (ISS) in 2013. His brain had adjusted to a space environment.

After landing back on Earth, with the full force of gravity now pulling on Hadfield, he became off balance and sick to his stomach. He had trouble walking. His inner ears, eyes, and brain had to re-adapt to sense which way was up. And even though he had exercised two hours per day on the ISS, his muscles weakened and his bones became more brittle. It took weeks for him to feel better.

NASA scientists fear that when astronauts land on Mars after spending eight months in weightlessness, they will be too sick to perform their mission. Mars has about one-third the gravity of Earth. If astronauts need to spend time in their landing craft getting used to gravity, more air, food, water, and power will need to be brought along. It could be fatal if a disoriented astronaut walked outside too soon and fell, breaking a bone or suffering a concussion.

Another danger that astronauts will face is radiation. Mars does not have a protective magnetic field surrounding the planet, like Earth. Hazardous cosmic rays and radiation from the Sun constantly bombard the Martian surface. They can harm brain cells, cause radiation sickness, and cause cancer later in life.

Spacecraft and habitats can be shielded with lead, but that makes them too heavy. Walls could possibly be filled with water, which also shields against radiation. Or maybe a device will be invented that makes an artificial magnetic field. New drugs might also help the human body fight off the effects of radiation.

Astronauts Scott and Mark Kelly are identical twins. In 2015 and 2016, Scott spent 340 days aboard the International Space Station, while Mark stayed on Earth. The twins had many medical tests before, during, and after the mission. Identical twins share nearly the same genetic material. Any differences in the test results would likely be from the effects of living in space. NASA especially wanted to know how the human body reacts to weightlessness, radiation, and the mental stress of isolation.

Astronauts Scott (left) and Mark Kelly.

Besides coming back to Earth two inches (5 cm) taller than his brother (a lack of gravity causes spinal disks to expand—they eventually shrink back to normal), some of Scott Kelly's genes changed. They affected his body's immune system, how his bones form, and how his DNA repairs itself. Many of the changes went back to normal when Kelly returned to Earth, but not all. When doctors finish studying the medical test results, they hope to have a better understanding of the health of future Mars explorers.

CHAPTER 77
SPACESUITS

A spacesuit is like a wearable spacecraft. It comes with its own air supply, heating and cooling systems, and radio. Spacesuits used by Martian explorers will differ from those used on the space shuttle or the International Space Station, or even those used by the moonwalkers of the Apollo missions.

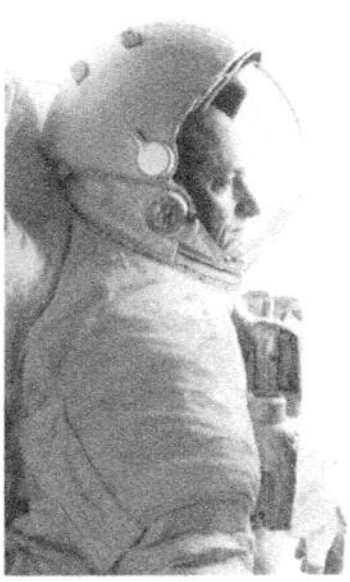

On the hostile surface of Mars, a spacesuit will need a life-support system that lasts long enough for astronauts to complete tasks and experiments that might last for several hours. The suits must be designed for walking on the rocky surface, which means they have to be tough, light, and flexible in the joints, especially the knees, waist, and ankles.

Since the spacesuits used on Mars will be worn for many months, they will need to be very durable. Nobody is sure what fabrics will last the longest in the Martian environment. The Mars 2020 rover will carry fabric samples, including swatches of Teflon and polyurethanes, to see how they wear compared to their durability on Earth.

The design of Martian spacesuits could take many forms. It is likely that at first they will look like the bulky, rigid balloon-shaped suits currently used. They look this way because they are filled with pressurized air to simulate the atmospheric pressure on Earth. Other suits might include a more skintight design. These elastic suits apply pressure directly to the skin. Cables and coils "shrink wrap" astronauts into the suit. They would help astronauts be more flexible while they work. These types of suits are still experimental today, but they may be commonplace by the time humans are sent to Mars sometime around 2040.

CHAPTER 78
FINDING WATER

For a yearlong stay on Mars, a crew of astronauts will probably need more water than they can carry from Earth. They will have to use resources found on Mars to survive. Scientists call this "in-situ resource utilization." In situ means "in its original, natural place."

In the past, scientists believed Mars was a dry, lifeless desert. We now know, thanks to a fleet of Mars orbiters and rovers, that the planet has huge amounts of water. Because of Mars's low atmospheric pressure, liquid water on the surface quickly evaporates. However, there is evidence of vast stores of frozen water under the Martian soil. The polar regions, especially the north pole, are blanketed with water ice. In fact, if all the ice were to suddenly melt, it would flood the entire planet with water 33 feet (10 m) deep.

Getting to the water ice and using it will require special equipment. Microwaves could be used to heat up the soil, and the water vapor could be collected. Water vapor in small amounts is also found naturally in the Martian atmosphere. It is possible to use absorbent materials to capture this water.

CHAPTER 79
GROWING FOOD

ASTRONAUTS WILL HAVE TO GROW THEIR OWN FOOD DURING A LONG stay on Mars. They probably cannot bring enough freeze-dried food with them on the landing ship. For a mission that could last more than a year, fresh food will be important for proper nutrition and variety. Also, plants grown on Mars will convert carbon dioxide to oxygen, just like on Earth.

Growing crops on Mars isn't as simple as it seems. Rays from the Sun are weaker because Mars is millions of miles farther away than Earth. Also, there are chemical differences between Martian and Earth soils. Plants need essential nutrients to grow, including the right amounts of phosphorus, iron, potassium, and nitrogen. These problems might be overcome by using LED light banks inside greenhouses, and by using fertilizers to treat the Martian soil.

The crew of a Mars base will probably include someone who knows a lot about botany and biology. In the movie *The Martian*, marooned astronaut Mark Watney was a botanist. His skills helped him survive by growing potatoes in the Martian soil.

Are potatoes really the best crop to grow on Mars? NASA is performing tests aboard the International Space Station to see what kinds of plants can thrive in low gravity and in soil that isn't as fertile as on Earth. Plants such as tomatoes, rye, carrots, and certain herbs, including garden cress, have shown promise.

CHAPTER 80
MAKING FUEL

WEIGHT IS A CRITICAL PART OF SPACE TRAVEL. HEAVIER LOADS NEED more fuel to get to Mars. Fuel itself is heavy, which makes things even more complicated. A heavy spacecraft has to be big just to carry the fuel that will be burned in order to safely land on the surface. That doesn't leave much room for the fuel needed to return to Earth.

If we could somehow make rocket fuel on Mars, we wouldn't have to carry as much, and spaceships could be smaller and lighter. Luckily, Mars has chemicals in abundance that can be converted to rocket fuel.

Mars has frozen water at the poles and underground. The planet's atmosphere contains mostly carbon dioxide. These two resources, when converted by special equipment, can produce oxygen to breathe and rocket fuel.

The first step is to find water ice, melt it, and then run an electric current through it. This process is called electrolysis. The electricity can be provided by solar arrays. The process releases hydrogen for fuel to burn, and oxygen to breathe. Also, hydrogen and oxygen can be combined and used as rocket fuel.

By mixing carbon dioxide in the Martian atmosphere with the hydrogen created by electrolysis, astronauts can create drinking water and methane gas. The methane can be used as rocket fuel. This chemical reaction is called the Sabatier process. It is named after Paul Sabatier, the French chemist who invented it in 1910.

By using these simple chemical reactions, much of the material needed by future Mars explorers will not need to be brought by spacecraft. It already exists, right there on the planet.

CHAPTER 81
TERRAFORMING MARS

AFTER EXPLORING MARS AND BUILDING A LARGE COLONY, THE NEXT step might be to make Mars more like Earth. Right now, if an astronaut were to walk on the surface of Mars without a spacesuit, his or her exposed body fluids, including the linings of the lungs, would start boiling. The atmospheric pressure is too low. Also, there isn't enough oxygen in the Martian air for humans to breathe. Suffocating while one's bodily fluids boil away would be quite unpleasant. Not to mention skin and eyeballs freezing solid when nighttime temperatures drop well below -100 degrees Fahrenheit (-73° C).

Even though the Martian environment is hostile to humans now, there are good reasons for creating a place other than Earth where people can live. What if a catastrophe wiped out most life on Earth? We know it happened once before when an asteroid killed the dinosaurs. Other disasters might include nuclear war, an incurable virus, or runaway global warming. To avoid humanity's extinction, it might be a good idea to have someplace, a Planet B, where brave pioneers can keep the human race alive.

Of all the planets in our solar system, Mars is the best candidate for being turned into a habitable world like Earth. This process is called terraforming. Right now we don't have all the technology we need to terraform Mars, but that is only a matter of time and determination.

How might future astronauts terraform Mars? One plan is to start by increasing the planet's atmospheric pressure. This is important so that liquid water can exist on the surface. There are huge amounts of frozen carbon dioxide at the poles, especially the Martian south pole. During the summer, some of the carbon dioxide ice heats up and turns to gas—a greenhouse gas that thickens the atmosphere and makes the planet warmer. If we could somehow heat all the polar carbon dioxide ice—perhaps with giant solar mirrors in orbit above the planet—it might make the atmosphere warm and thick enough to support water on the surface. It would take a few centuries, but water ice at the poles and underground would start to flow, flooding the surface with oceans and lakes.

Next, humans on Mars would need more oxygen to breathe. Making the soil fertile for plants would help, since plants naturally produce oxygen. Microbes and mosses from Earth, or grown in a lab, could cover the Martian surface. Other plants and trees would follow in the decades to come. Mars would be transformed into a wet, green planet. It would not be a second Earth, but humans could thrive there.

There are many obstacles to exploring Mars. But one thing seems certain, whether we terraform it or not: we will always be fascinated by the Red Planet. It calls to us, that blood-red speck in the night sky. And if we learn enough by studying our sister planet's history, maybe we can use that knowledge to protect and care for the little blue ball in space we call home: Earth.

ALSO IN THE DESTINATION OUTER SPACE SERIES

Join award-winning science writer John Hamilton as he traces humanity's race to the moon and the exploration of the final frontier.

The Space Race tells the story of human space exploration, from the dawn of rocketry to the Apollo Moon landings. Every crewed American space mission is included, from Project Mercury to Gemini to Apollo. Also covered are vital details about important Soviet spaceflights, as well as America's space shuttle, the International Space Station, and planned future missions.

Why did NASA fear that astronaut John Glenn would burn up on reentry during his historic Friendship 7 spaceflight? Who was the first person to vomit in space? Why did Neil Armstrong almost crash the Eagle lunar lander before becoming the first person to walk on the Moon? These questions and many more are answered in these inspiring—and sometimes humorous—stories of triumph, tragedy, and determination. Perfect for school reports, but also a thrilling book for anyone, young or old, interested in space exploration. ***"... readers will find themselves immersed in the excitement."*****—American Library Association Booklist.**

In this 36,000-word history of the space race, you'll discover:

• Early rocketry, from Chinese fireworks to the terrifying German V-2 missiles of WWII.

• The Cold War rivalry between the United States and the Soviet Union.

• The origins of NASA.

• The original Mercury 7 astronauts.

• The cosmonauts who went where no one had gone before.

• The spacecraft that brought us to the Moon.

• Concise biographies of the pioneering men and women who first explored the heavens.

• A helpful glossary and timeline.

• Aligned to Common Core Standards and correlated to state standards.

• Over 75 historical images from NASA's archives.

• And much, much more!

The Space Race is thoroughly researched, easy to read, and filled with adventure. Boldly seek out your copy today!

Click here for more information on Amazon Kindle: *The Space Race*

Or scan the QR code below with your smart device:

GET A FREE BOOK

If you enjoyed this book, join my readers' group and get a free copy of *Worlds Above*, an exploration of the solar system. My newsletter is how I keep my loyal readers up-to-date about my latest releases, and send news about my writing and research trips. Click below to get started.

Get my free book

Or scan the QR code below with your smart device:

PLEASE LEAVE A REVIEW

If you enjoyed *Mission: Mars,* please consider leaving a review. Reviews are the lifeblood of independent authors. They help new readers who might be on the fence. They also help me discover what's working and what could use improvement. I do so enjoy hearing from my readers. It would mean a lot to me if you would add a line or two, or even simply add a star rating. Thanks, and happy reading!

Go directly to the Amazon review page by clicking the following link: Mission: Mars

Or scan the QR code below with your smart device:

TIMELINE

Mission: Mars

2560 bc—The Great Pyramid of Giza, also known as the Pyramid of Cheops, finishes construction by the Egyptians. Its north-south axis is aligned with the North Star.

2500 bc—Prehistoric people in today's England finish construction of Stonehenge, a collection of huge standing stones arranged in a circle. The site may have been used as a burial ground, and to predict the position of the Sun and Moon.

500 bc—Classical Greek civilization begins. The ancient Greeks believe in many gods, including Ares, the god of war. The ancient Romans later called Ares as we know him today—Mars.

1877—Italian astronomer Giovanni Schiaparelli draws first detailed map of Mars. He sees long, straight lines on the planet, which he calls canali.

1877—American astronomer Asaph Hall discovers the two Martian moons, Phobos and Deimos.

1895-1908—American astronomer Percival Lowell publishes a series of books and includes detailed drawings of the Martian "canals" (which later prove to be optical illusions). Lowell's books convince many people that intelligent life resides on Mars.

1897—*The War of the Worlds,* by English author H.G. Wells, first published.

1917—*A Princess of Mars,* by American author Edgar Rice Burroughs, first published.

Oct. 30, 1938—Orson Welles's *The War of the Worlds* radio broadcast causes widespread panic.

1951—*The Martian Chronicles,* by American author Ray Bradbury, published.

July 14-15, 1965—Mariner 4 (USA) spacecraft, first successful flyby of Mars. The spacecraft takes 22 low-resolution photos of Mars, revealing a dry, rocky planet filled with Moon-like craters.

July 30, 1969—Mariner 6 (USA) flyby. Comes within 2,136 miles (3,437 km) of Martian equatorial surface.

Aug. 4, 1969—Mariner 7 (USA) flyby. Records temperature at south pole of -190° Fahrenheit (-123° C).

Nov. 14, 1971—Mariner 9 (USA) orbiter arrives at Mars. First United States spacecraft to orbit a planet other than Earth.

Nov. 27, 1971—The Soviet Union's Mars 2 orbiter arrives at Mars. Lander crashes, becoming the first human object to reach the surface of planet.

Dec. 2, 1971—Mars 3 (USSR) orbiter arrives at Mars. Along with sister spacecraft Mars 2, automatically takes photos as featureless dust storm swirls on planet. Mars 3 lander touches down on surface but malfunctions after a few seconds. Prop-M rover never has a chance to deploy.

June 19, 1976—NASA's Viking 1 orbiter arrives at Mars and takes images of the surface, including the strange, face-like rock formation on the Martian region of Cydonia.

July 20, 1976—Viking 1 (USA) lander touches down on Mars. First lander to safely land on Mars and complete its mission of photographing the Martian landscape and analyzing soil samples.

Aug. 7, 1976—Viking 2 (USA) orbiter arrives at Mars. Together with Viking 1, the two orbiters send nearly 52,000 color images of the Martian surface NASA scientists on Earth.

Sept. 3, 1976—Viking 2 (USA) lander touches down on the surface of Mars.

1988—PBS first broadcasts *The Power of Myth*, a series that features Joseph Campbell, the writer who popularized character archetypes and mythology.

July 4, 1997—Mars Pathfinder (USA) safely lands. Monitors weather and takes high-resolution photos of surface. Accompanying Sojourner rover successfully deployed. Lander renamed Carl Sagan Memorial Station.

Sept. 11, 1997—Mars Global Surveyor (USA) arrives in polar orbit around Mars using aerobraking to slow its speed. Over nine years, it takes thousands of photos of Mars and its two moons.

September 27, 1997—Radio communications from Pathfinder and Sojourner (USA) cease. They probably fell victim to the extreme cold temperatures on Mars.

Dec. 3, 1999—Mars Polar Lander crashes on surface because retrorockets shut down too soon.

Oct. 24, 2001—2001 Mars Odyssey (USA) enters Mars orbit. Discovers massive amounts of water ice under the Martian surface. Also studies Mars's geology and radiation, and acts as communications relay for Martian rovers. It is the longest-serving Mars orbiter to date.

Dec. 25, 2003—The European Space Agency's Mars Express orbiter arrives at the Red Planet. Accompanying Beagle 2 lander begins its descent to the Martian surface. Attempts to contact the spacecraft after landing fail. Mars orbiter images later show that Beagle 2's solar panels fail to properly unfold. The Beagle 2 lander's mission fails, but Mars Express orbiter is successful. Explores geology and atmosphere of the planet, takes high-resolution stereo images.

Jan. 4, 2004—Spirit (USA) rover lands safely on Mars.

Jan. 25, 2004—Opportunity (USA) rover lands safely on Mars.

March 10, 2006—Mars Reconnaissance Orbiter (USA) begins orbiting Mars. Telescope camera takes high-resolution images of planet's surface. Other instruments explore different minerals on

Mars, and water ice trapped under surface. Acts as radio relay for other spacecraft on surface or in orbit around Mars.

May 25, 2008—Phoenix Mars Lander (USA) safely touches down in Martian north polar region. First lander to make direct contact with water ice on Mars.

March 22, 2010—Spirit sends its last radio signals to Earth. After being stuck in deep sand for months, it probably succumbed to the harsh Martian weather conditions.

2011—*The Martian,* by America author Andy Weir, published.

August 6, 2012—The Mars Science Laboratory (USA) mission's rover, Curiosity, touches down on Mars using a unique and daring "sky crane" landing technique.

Sept. 22, 2014—MAVEN (USA) orbiter enters orbit around Mars. Investigates how the Martian upper atmosphere is affected by the solar wind.

Sept. 24, 2014—Mars Orbiter Mission (India), also called *Mangalyaan* (Hindi for "Mars Craft"), reaches Mars and begins studying the planet's atmosphere and surface details. It is India's first mission to Mars.

Oct. 19, 2016—ExoMars Trace Gas Orbiter (joint European Space Agency and Roscosmos (Russia)) enters orbit around Mars. The ExoMars lander, Schiaparelli EDM (Entry, descent and landing Demonstrator Module), crashes into Mars. Retrorockets shut down too soon, causing spacecraft to hit surface at high speed, destroying it.

June 2018—NASA announces that Curiosity has discovered organic molecules inside ancient Martian rocks, and methane in the air that increases in the warmer summer months. Both may be signs of microbial life on Mars, but further investigation is needed.

Nov. 26, 2018—InSight (USA) lander safely touches down. Using a seismometer and other instruments, it studies the deep interior of Mars. Two small CubeSat spacecraft accompany the lander, monitoring the landing and relaying communications to Earth.

Feb. 13, 2019—NASA declares Opportunity's mission over, after months of not being able to contact the rover due to a dust storm that enveloped Mars in June 2018.

February 18, 2021—NASA's Perseverance rover lands safely in Jezero Crater and begins searching for clues to possible microbial life in Mars's ancient, watery past.

May 14, 2021—China's Zhurong rover lands in Mars's Utopia Planitia region.

GLOSSARY

Abolitionist

An activist who fought for freedom for slaves.

Asteroid

A rocky object, smaller than a planet, that revolves around the Sun, usually between the orbits of Mars and Jupiter. Their size ranges from one to several hundred miles in diameter. Mars's two moons, Phobos and Deimos, are probably asteroids captured by the planet's gravitational pull millions of years ago.

Columbia Hills

A range of small hills inside Gusev Crater. The Spirit rover explored the area for several years during its mission. The range is named after the destroyed space shuttle Columbia. The seven peaks are named after the astronauts who died during the disaster in 2003.

Cosmic Rays

High-energy radiation that usually comes from outside the Solar System, possibly from the supernova of stars in our galaxy, or even from other galaxies. Without the protection of Earth's atmosphere, astronauts that are exposed to this radiation can be stricken with cancer.

Cult

A religious group that especially worships a special person, goal, or even an object. In ancient Roman culture, the cult of Mars became popular as the Roman army invaded other lands.

European Space Agency (ESA)

A space agency, like NASA, that builds and flies spacecraft that explore the solar system. Its headquarters is in Paris, France. As of 2019, there are 22 countries that are members of the ESA.

Extraterrestrial Life

Life that does not come from the planet Earth, popularly called alien life. It might range from simple germs to complex, intelligent organisms. So far, extraterrestrial life has not yet been discovered. Many experiments on Mars have been conducted in order to find out if extraterrestrial life may once have lived on the Red Planet.

Flyby

When a spacecraft travels close to a planet or other object but does not enter into an orbit around it. During a flyby, a spacecraft has one chance to take as many photos and gather as much scientific data as possible before it sails off to its next destination.

Gale Crater

Gale Crater is the large crater where the Curiosity rover landed. It is named after Walter Frederick Gale (1865-1945), an Australian amateur astronomer who studied Mars in the late 1800s. He also discovered several new comets.

Gusev Crater

A 103-mile (166-km) -diameter crater in the Aeolis region of Mars. Explored by the Spirit rover, Gusev Crater is named after

Russian astronomer Matvei Gusev (1826-1866), who was a pioneer of astrophotography.

Interplanetary

Located or traveling between the planets of a solar system.

Jet Propulsion Laboratory (JPL)

A NASA-owned research center based in Pasadena, California. It is managed for NASA by the California Institute of Technology (Caltech). JPL's main mission is to plan and construct robotic spacecraft that explore space and our solar system, including Mars.

Meteoroid

A solid object, usually rocky, that moves through space. It is much smaller than an asteroid, ranging from the size of a grain of sand to about one meter in diameter. If a meteoroid enters a planet's atmosphere, it becomes a meteor. Earth's atmosphere is so thick that most meteors burn up, becoming streaks in the sky, which we call "falling" or "shooting stars." If a meteor survives and strikes the ground, it is called a meteorite.

Microbe

Small life-forms (microorganisms), such as bacteria, fungi, or viruses. The Curiosity rover's main mission was to explore whether Mars could be home to microbial life in its distant past.

National Aeronautics and Space Administration (NASA)

A United States government space agency started in 1958. NASA's goals include space exploration, as well as increasing people's understanding of Earth, our solar system, and the universe.

National Space Policy

A set of rules and goals for the space program of the United

States. It is set by the president, with advice from science and industry experts. It is then approved and funded by Congress.

Olympians

Major gods of ancient Greek culture. They lived atop Mount Olympus. The 12 Olympians included Aphrodite, Apollo, Ares, Artemis, Athena, Demeter, Hestia, Hephaestus, Hera, Hermes, Poseidon, and Zeus. Ares, the god of war, would later come to be called Mars by the ancient Romans.

Orbit

The circular path a moon or spacecraft makes when traveling around a planet or other large celestial body. There are several satellites orbiting Mars, including NASA's Mars Reconnaissance Orbiter and the European Space Agency's ExoMars Trace Gas Orbiter.

Probe

An unmanned space vehicle that is sent on missions that are too dangerous, or would take too long, for human astronauts to accomplish. Probes are equipped with many scientific instruments, like cameras and radiation detectors. Information from these instruments is radioed back to ground controllers on Earth.

Roscosmos (Roscosmos State Corporation for Space Activities)

A government agency in charge of space flight missions for the Russian Federation. Headquartered in Moscow, Russia, it was established on February 25, 1992. It was formerly known simply as the Soviet space agency, a program of the former Soviet Union.

Rover

A robotic vehicle that is driven over rough terrain by remote control.

Sedimentary

Rock that is formed by sediments (tiny particles of rocks carried by water) that collect and press together over thousands or millions of years. Sandstone is a sedimentary rock.

Solar Wind

Streams of charged particles that are given off by stars. Solar wind is a plasma of electrons, protons, and other particles. They are so energetic they can escape from the Sun's gravity.

Soviet Union

A former country that included a union of Russia and several other communist republics. It was formed in 1922 and existed until 1991.

Space Race

The Space Race was a competition between the United States and the former Soviet Union (much of which is today's Russia). It started in the mid-1950s and lasted until the early 1970s. By the mid-1960s, both countries had sent probes to Venus. Mars was the next prize.

Spectrometer

A spectrometer is a recording device that lets scientists collect data about light (the electromagnetic spectrum) that humans can't see, such as gamma rays, x-rays, microwaves, and radio waves. Light that humans can see is called "visible light." It is just a small part of the entire electromagnetic spectrum.

Telescope

A device to detect and observe distant objects by their reflection or emission of various kinds of electromagnetic radiation (like light). Most astronomy instruments today detect electromagnetic radiation other than visible light, such as radio or x-ray telescopes.

Terraform

Modifying a planet on purpose to make its atmosphere, temperature, and ecology similar to Earth's. The dream of some scientists is to terraform Mars so that large populations of colonists from Earth can someday live on the Red Planet.

SELECTED BIBLIOGRAPHY

- Achenbach, Joel. "Mars: The Race to the Red Planet." *National Geographic*, vol. 230, no. 5. November 2016, pp. 31-61.
- Aldrin, Buzz, et al. *Mission to Mars*. Time Inc., 2016.
- "Big Fans." NASA, 8 Mar. 2018, www.nasa.gov/image-feature/jpl/pia22332/big-fans.
- "By the Numbers | Mars – NASA Solar System Exploration." NASA Solar System Exploration, solarsystem.nasa.gov/planets/mars/by-the-numbers/.
- Chua, Jasmin M. "Here's What Spacesuits for Travel to Mars Might Look Like." Racked, 18 June 2018, www.racked.com/2018/6/18/17466150/mars-spacesuit.
- "Description of the Rover Sojourner." NASA's Mars Exploration Program, mars.nasa.gov/MPF/rover/descrip.html.
- "Dry Ice Snowfall at the Poles of Mars." The Planetary Society, 16 Jan. 2014, www.planetary.org/articles/20140116-dry-ice-snowfall-at-the-poles-of-mars.

- "ESA Tests Its Giant Mars Mission Parachute." Engadget, 31 Mar. 2018, www.engadget.com/2018-03-31-esa-tests-giant-exomars-parachute.html.
- "Geologic History Revealed in Valles Marineris." NASA, 7 Feb. 2018, www.nasa.gov/image-feature/jpl/pi-a22238/geologic-history-revealed-in-valles-marineris.
- Hadhazy, Adam. "How Fast Could Humans Travel Safely Through Space?" BBCpage, www.bbc.com/future/article/20150809-how-fast-could-humans-travel-safely-through-space.
- "An Introduction to Mars." Planetary Sciences, Inc. – – Inspiring the Next Generation of Space Explorers –, planetary-science.org/mars-research/mars/.
- "Life on Mars: Exploration & Evidence." Space.com, 15 Dec. 2017, www.space.com/17135-life-on-mars.html.
- "Mars Close Approach | Mars in Our Night Sky – NASA's Mars Exploration Program." NASA's Mars Exploration Program, mars.nasa.gov/allabout-mars/nightsky/mars-close-approach/.
- "Mars Cube One (MarCO)." NASA Jet Propulsion Laboratory (JPL) - Robotic Space Exploration, www.jpl.nasa.gov/missions/mars-cube-one-marco.
- "Mars Exploration: Multimedia." NASA's Mars Exploration Program, mars.nasa.gov/gallery/atlas/olympus-mons.html.
- "Mars Facts | All About Mars – NASA's Mars Exploration Program." NASA's Mars Exploration Program, mars.nasa.gov/all-about-mars/facts/#?c=inspace&s=distance.
- "Mars Helicopter to Fly on NASA's Next Red Planet Rover Mission." NASA, 11 May 2018, www.nasa.gov/press-release/mars-helicopter-to-fly-on-nasa-s-next-red-planet-rover-mission.
- "Mars in a Minute: How Do You Get to Mars?" NASA/JPL Edu, 31 Dec. 2020, www.jpl.nasa.-

gov / edu / learn / video / mars-in-a-minute-how-do-you-get-to-mars / .

- "Mars: Crash Course Astronomy #15." YouTube, www.youtube.com / watch?v=I-88YWx71gE.
- "The Martian Mole Is At Last Fully Underground, But The Next Step is Critical." IFLScience, 4 June 2020, www.iflscience.com / space / the-martian-mole-is-at-last-fully-underground-but-the-next-step-is-critical / ? fbclid=IwAR3Fa7ORSYHk2QEChx0OnYjeyrUK4xPz6f-RHIpKlav42dnanZf6N-2BBkn8.
- "Meteorite Impacts Expose Ice on Mars." Science Mission Directorate | Science, 24 2009, science.nasa.-gov / science-news / science-at-nasa / 2009 / 24sep_martianice.
- "Mission Updates | Mission – NASA's Mars Exploration Program." NASA's Mars Exploration Program, mars.nasa.gov / msl / mission-updates / .
- "Moon to Mars Overview." NASA, 29 June 2018, www.nasa.gov / topics / moon-to-mars / overview.
- "NASA - NSSDCA - Spacecraft - Details." Welcome to the NSSDCA, nssdc.gsfc.nasa.gov / nmc / spacecraft / dis-play.action?id=1971-051A.
- "NASA Selects Developer for Rocket to Retrieve First Samples from Mars." NASA, 7 Feb. 2022, www.nasa.gov / press-release / nasa-selects-developer-for-rocket-to-retrieve-first-samples-from-mars.
- "NASA's James Webb Space Telescope to Reveal Secrets of the Red Planet." NASA, 20 Feb. 2018, www.nasa.gov / feature / goddard / 2018 / nasas-james-webb-space-telescope-to-reveal-secrets-of-the-red-planet.
- "NASA's Odyssey Orbiter Marks 20 Historic Years of Mapping Mars." NASA, 7 April 2021, https: / / mars.-nasa.gov / news / 8911 / nasas-odyssey-orbiter-marks-20-historic-years-of-mapping-mars /

- "New Mars Gravity Map." NASA, 16 Mar. 2016, www.nasa.gov / feature / goddard / 2016 / mars-gravity-map.
- "NextSTEP Partners Develop Deep Space Habitat Ground Prototypes." NASA, 9 Aug. 2016, www.nasa.gov / feature / nextstep-partnerships-develop-ground-prototypes.
- "Olympus Mons From Orbit." Astronomy Picture of the Day, apod.nasa.gov / apod / ap981019.html.
- "Overview | Mission – NASA's Mars Exploration Program." NASA's Mars Exploration Program, mars.nasa.gov / msl / mission / overview / .
- "Phobos: Facts About the Doomed Martian Moon." Space.com, 8 Dec. 2017, www.space.com / 20346-phobos-moon.html.
- "Phobos | Martian Moons – NASA's Mars Exploration Program." NASA's Mars Exploration Program, mars.nasa.gov / all-about-mars / moons / phobos / .
- "Planetary Physical Parameters." Jet Propulsion Laboratory, ssd.jpl.nasa.gov / planets / phys_par.html.
- Pyle, Rod. *Mars: Making Contact*. Andre Deutsch, 2016.
- Redd, Nola T. "How Long Does It Take to Get to Mars?" Space.com, 14 Nov. 2017, www.space.com / 24701-how-long-does-it-take-to-get-to-mars.html.
- "Rover – NASA's Mars Exploration Program." NASA's Mars Exploration Program, mars.nasa.gov / msl / spacecraft / rover / summary / .
- "Scientists Shrink Chemistry Lab to Seek Evidence of Life on Mars." NASA, 13 Apr. 2018, www.nasa.gov / feature / goddard / 2018 / moma.
- "Seven Ways Mars InSight is Different." NASA, 21 Feb. 2018, www.nasa.gov / feature / jpl / seven-ways-mars-insight-is-different.

- "Space Launch System at a Glance." NASA, www.nasa.gov / sites / default / files / atoms / files / sls_at_a_glance_10202015.pdf.
- "The Speed with Which We Get to Mars." ESA Blog Navigator – Navigator Page for Active ESA Blogs, 18 Mar. 2016, blogs.esa.int / rocketscience / 2016 / 03 / 18 / the-speed-with-which-we-get-to-mars / .
- Strickland, Ashley. "Flows of 'water' on Mars May Actually Be Sand, New Study Reveals." CNN, 21 Nov. 2017, www.cnn.com / 2017 / 11 / 21 / world / mars-water-sand-rsl-study / index.html.
- "Technology." NASA's Mars Exploration Program, mars.nasa.gov / odyssey / mission / technology / .
- "Unmasking the Face on Mars." Science Mission Directorate | Science, 23 2001, science.nasa.gov / science-news / science-at-nasa / 2001 / ast24may_1.
- "Valles Marineris: Facts About the Grand Canyon of Mars." Space.com, 9 Dec. 2017, www.space.com / 20446-valles-marineris.html.
- "Valles Marineris: The Grand Canyon of Mars." NASA, 6 Mar. 2015, www.nasa.gov / multimedia / image-gallery / image_feature_83.html.
- "Water on Mars: Exploration & Evidence." Space.com, 18 Aug. 2018, www.space.com / 17048-water-on-mars.html.
- "What Color is the Sky on Mars?" Serious Science, serious-science.org / what-color-is-the-sky-on-mars-7310.
- "Why Alien Life Would be our Doom - The Great Filter." YouTube, www.youtube.com / watch?v=UjtOGPJ0URM.
- "Witness the First Interplanetary Launch from West Coast." NASA, 12 Mar. 2018, www.nasa.gov / social / witness-the-first-interplanetary-launch-from-west-coast.

ABOUT THE AUTHOR

John C. Hamilton has spent most of his life in Minnesota. His first job was in the news media—delivering papers. He has worked as a house painter, a delivery driver, a corn shucker in a factory, and a movie usher learning the art of cinema, and life, in a darkened theater.

With a degree in photojournalism and mass communication in hand, he launched a career in magazine editing, which led to writing and editing books. After 40 years in the trenches, he is now a bestselling author of more than 300 nonfiction books for young adults, plus three novels and 13 screenplays, several of which have been optioned by Hollywood studios.

John's young adult book *Battle of the Little Bighorn* earned 2015 Spur Award Finalist honors from the Western Writers of America. He is a two-time winner of the Golden Duck Award for Excellence in Children's Science Fiction Literature, and his epic retelling of the Lewis & Clark expedition, *Lewis & Clark: Adventures West*, was a Minnesota Book Award finalist for Young Adult Nonfiction. *School Library Journal* said "(Hamilton's) books present a remarkable amount of information and provide readers with a clear understanding of complicated issues."

John can be found most summers with his family either exploring the American West or hiking along Minnesota's rugged North Shore.

Connect with John at:
www.johnchamilton.com
john@johnchamilton.com

Scan the QR code below with your smart device to go to John's website at www.johnchamilton.com:

ALSO BY JOHN C. HAMILTON

For the latest books and news, scan the QR code below with your smart device:

Heroes and Villains of the Wild West

Wild Bill Hickok

Wild Bill Hickok was a man of many skills: pioneer of the Great Plains, Civil War scout and spy, lawman, gambler, stagecoach driver, showman. He was a crack shot with nerves of steel. When he drew his weapon, it was with one aim: to kill. Wild Bill is brought to life in this easy-to-read title by award-winning Western author John Hamilton. *Wild Bill Hickok* is part of the bestselling Heroes and Villains of the Wild West hardcover series, now available for the first time as an ebook! Perfect for school reports, but a gripping tale for anyone captivated by the Wild West.

Jesse James

Jesse James was a villain to most people, a hero to others. He was a Missouri farmer, a ruthless Confederate guerrilla fighter, a loving husband and father, and a robber of banks, trains, and stagecoaches. He led a gang of desperadoes on a 16-year crime spree that sprawled over at least eight states. Jesse James is brought to life in this easy-to-read title by award-winning Western author John Hamilton. *Jesse James* is part of the bestselling Heroes and Villains of the Wild West hardcover series, now available for the first time as an ebook! Perfect for school reports, but a gripping tale for anyone captivated by the Wild West.

Other Nonfiction

Lewis & Clark: Adventures West

Originally published in hardcover and a Finalist at the 17th Annual Minnesota Book Awards, now available for the first time as an ebook! Perfect for school reports, but a gripping tale for everyone, young or old. Featuring 187 images; 4 maps; travel tips; glossary; index. *"An excellent stepping stone to in-depth adult titles such as Stephen E. Ambrose's Undaunted Courage."* — School Library Journal

Fiction

Ghost Marshal

All hell breaks loose when a badass woman meets a gunslinging ghost.

Jessie Parker is a woman on a mission: to avenge her father's brutal murder. But in 1876 Deadwood, she has about as much chance as a wax cat in hell—until she partners with the cantankerous ghost of Wild Bill Hickok. They soon find themselves knee-deep in gunfights, Chinese sorcery, barroom brawls, pleasures of the flesh, demon owls, forbidden romance, and a heaping dose of frontier justice. Along the way, Jessie and Bill uncover a bizarre conspiracy that takes them near enough to hell to smell smoke.

Isle Royale

Shipwrecks, gangsters, and the mother of all storms. Living in a lighthouse can be murder. How far would you go to be a hero?

Isle Royale is an adventure story filled with plot twists and compelling characters. If you like fast-paced excitement, danger on the high seas, and tales of bravery and redemption, then you'll love John Hamilton's page-turning thriller.

Made in United States
Cleveland, OH
18 October 2026

46341971R10136